T0311440

ARMIN SCHIEB

The Ant Collective

INSIDE THE WORLD OF AN ANT COLONY

Translated by Alexandra Bird

Princeton University Press
Princeton and Oxford

Contents

1.1 Anatomy of a worker

Most of the inhabitants of an anthill are wingless workers with a very uniform appearance. An important identifying feature is their red-brown-black coloring.

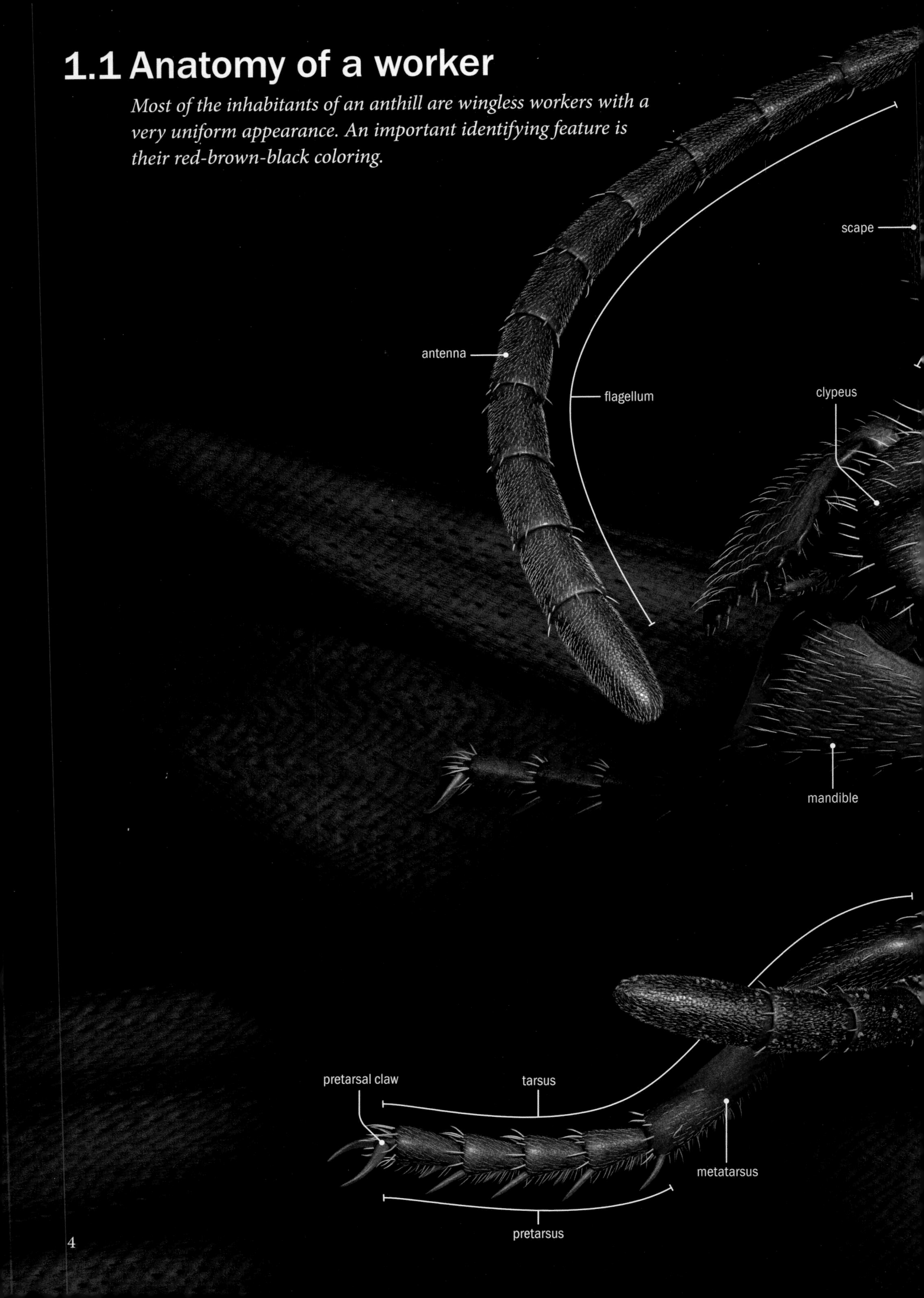

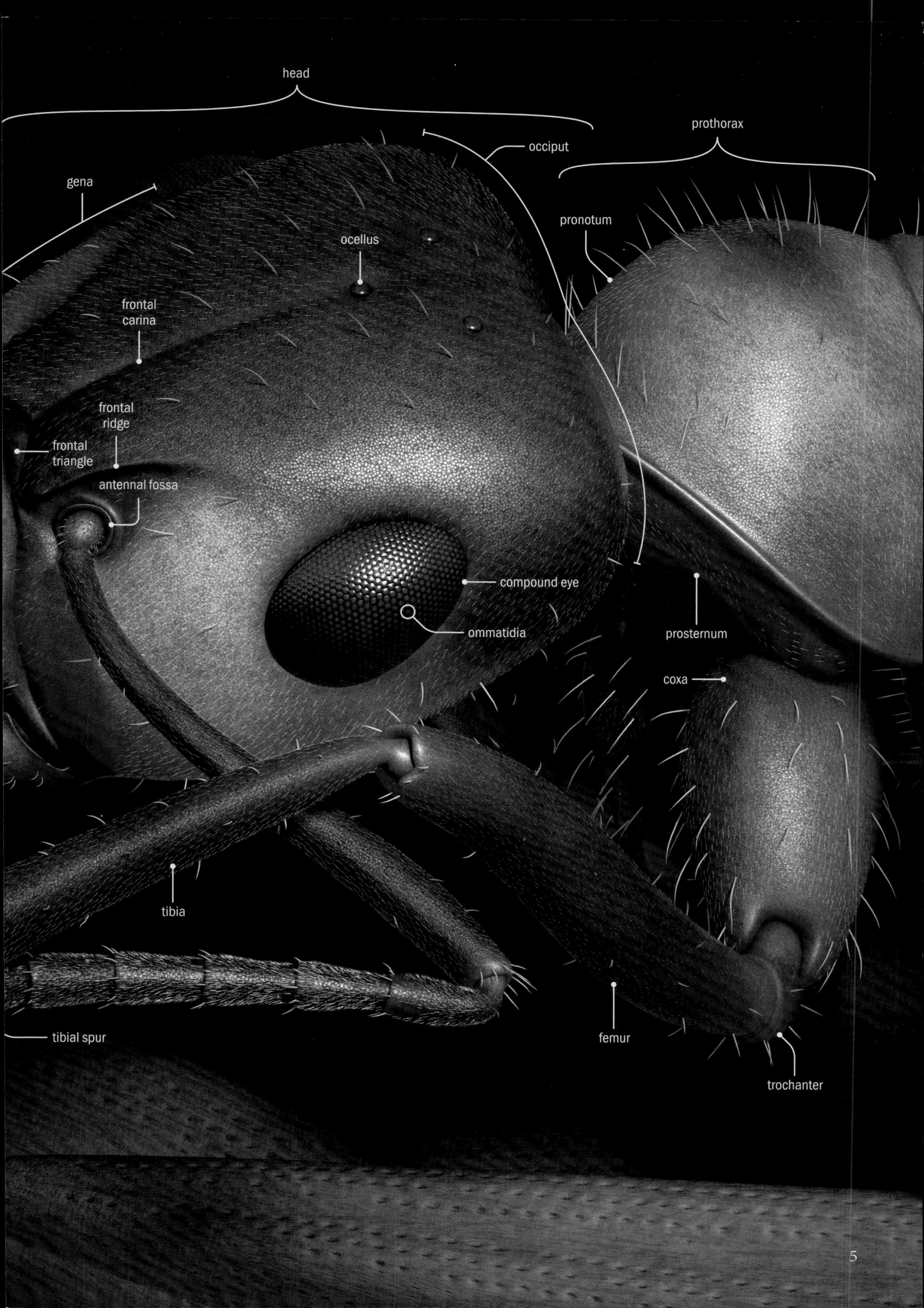
head
prothorax
occiput
gena
pronotum
ocellus
frontal carina
frontal ridge
frontal triangle
antennal fossa
compound eye
ommatidia
prosternum
coxa
tibia
tibial spur
femur
trochanter

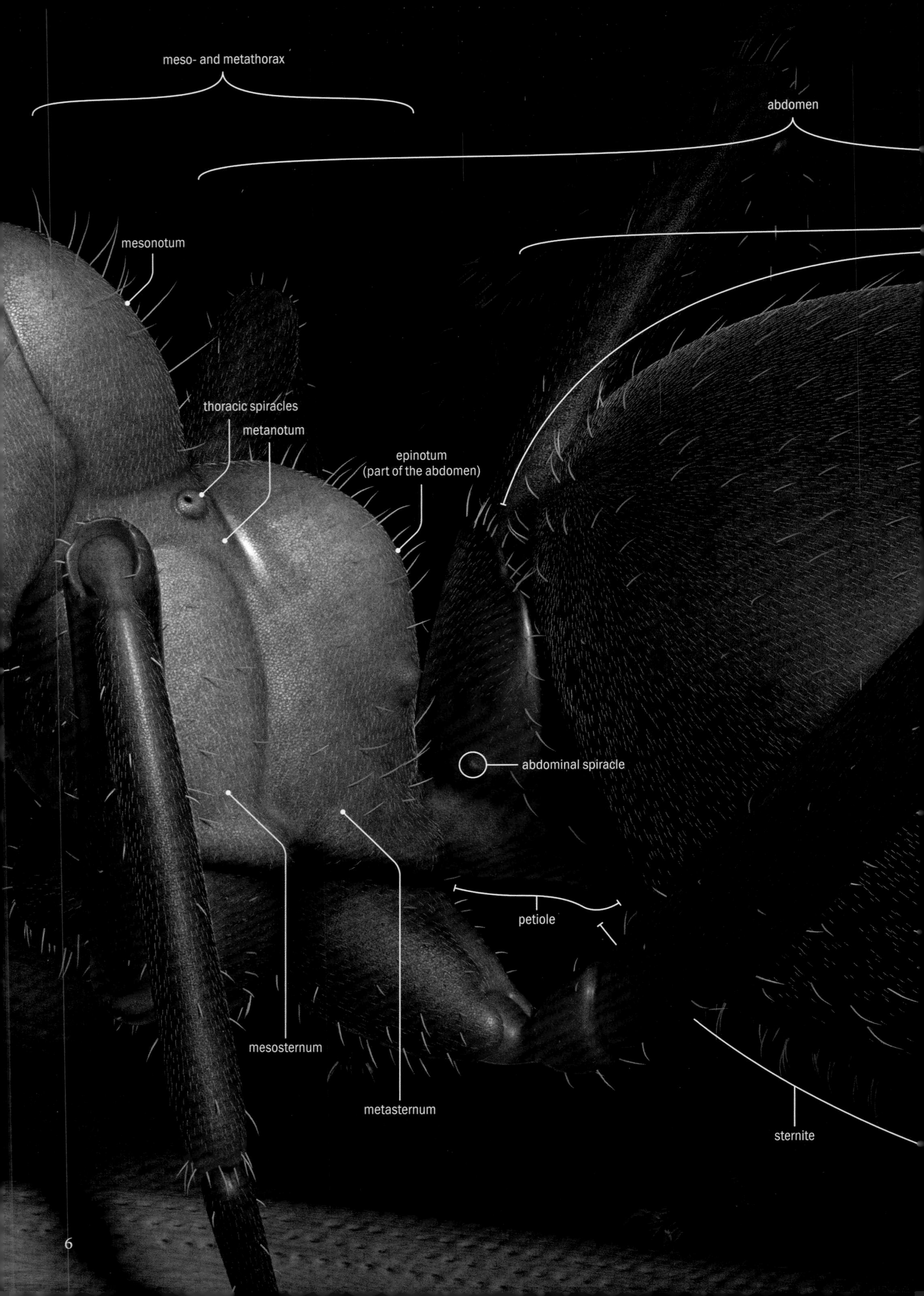
meso- and metathorax
abdomen
mesonotum
thoracic spiracles
metanotum
epinotum
(part of the abdomen)
abdominal spiracle
petiole
mesosternum
metasternum
sternite

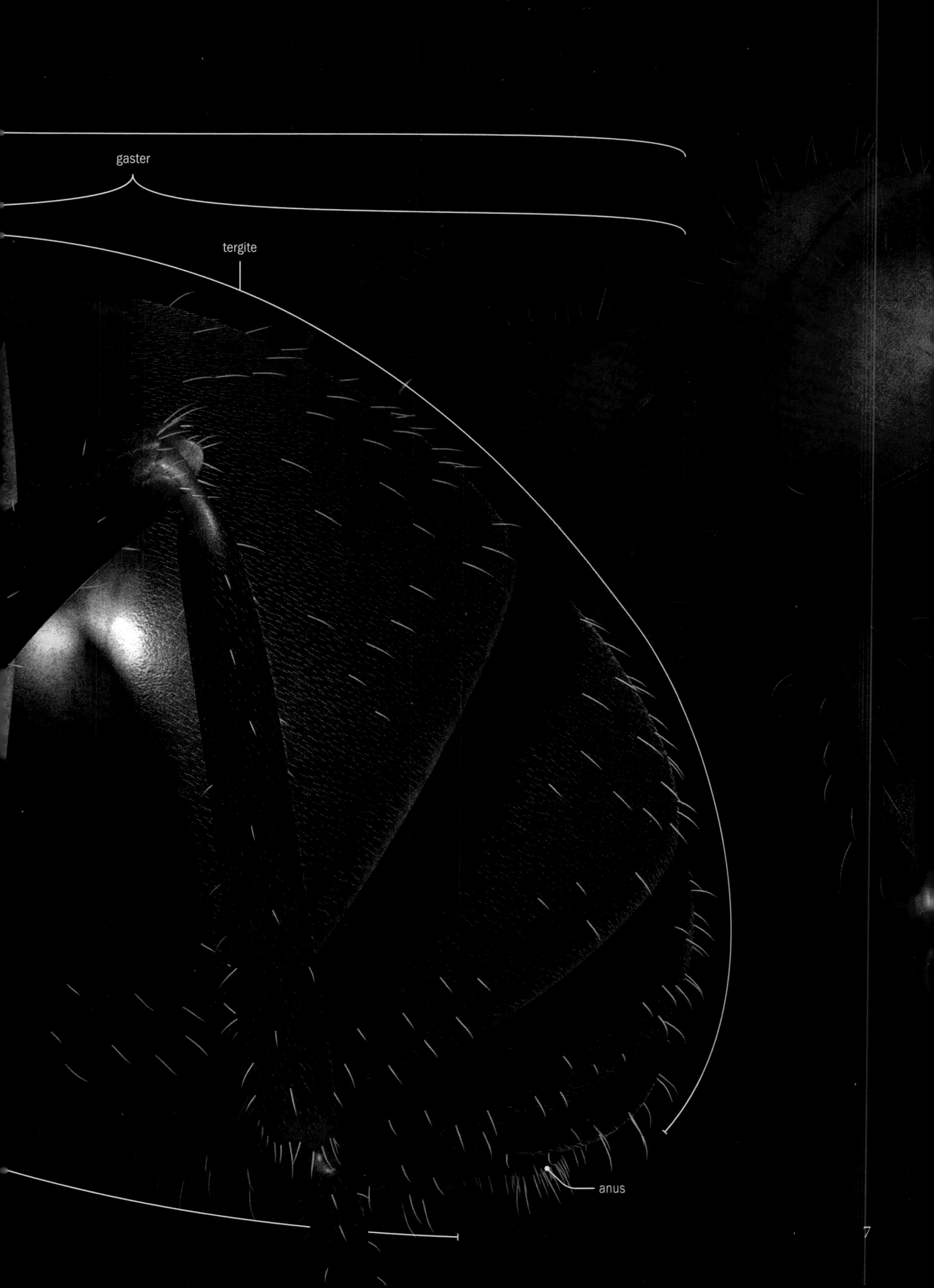
gaster
tergite
anus

1.2 Sex and caste

A red wood ant colony consists of three kinds of ants: males, female queens, and female workers. Each kind specializes in certain tasks.

4–9 mm

The numerous workers are the principal caste of the anthill. They perform all the tasks, such as caring for the larvae, hunting, defending the colony, or building the nest. Young worker ants limit themselves to tasks inside the nest; the older workers are responsible for the dangerous work outside. As workers don't normally reproduce, care of the egg-laying queen and larvae is their highest priority.

The males appear exclusively during mating season. Their sole function is the fertilization of young queens during the nuptial flight. They only take part in social life for a few days. They are unable to feed themselves and depend on the workers' care.

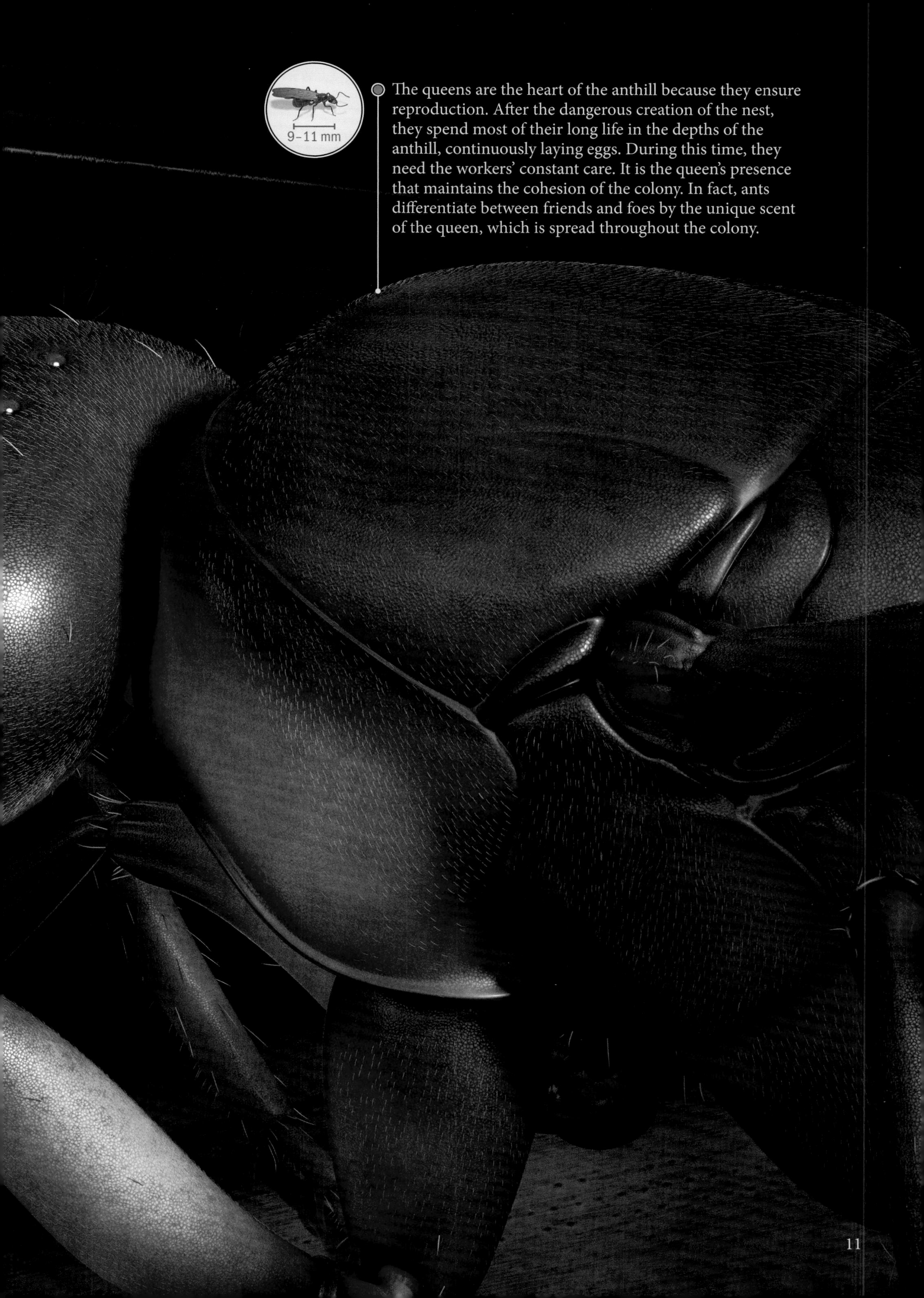

The queens are the heart of the anthill because they ensure reproduction. After the dangerous creation of the nest, they spend most of their long life in the depths of the anthill, continuously laying eggs. During this time, they need the workers' constant care. It is the queen's presence that maintains the cohesion of the colony. In fact, ants differentiate between friends and foes by the unique scent of the queen, which is spread throughout the colony.

1.3 Circulation and respiration

Ants have an open circulatory system to supply the organism with nutrients. The tracheae are a branching network of tubules that supply oxygen.

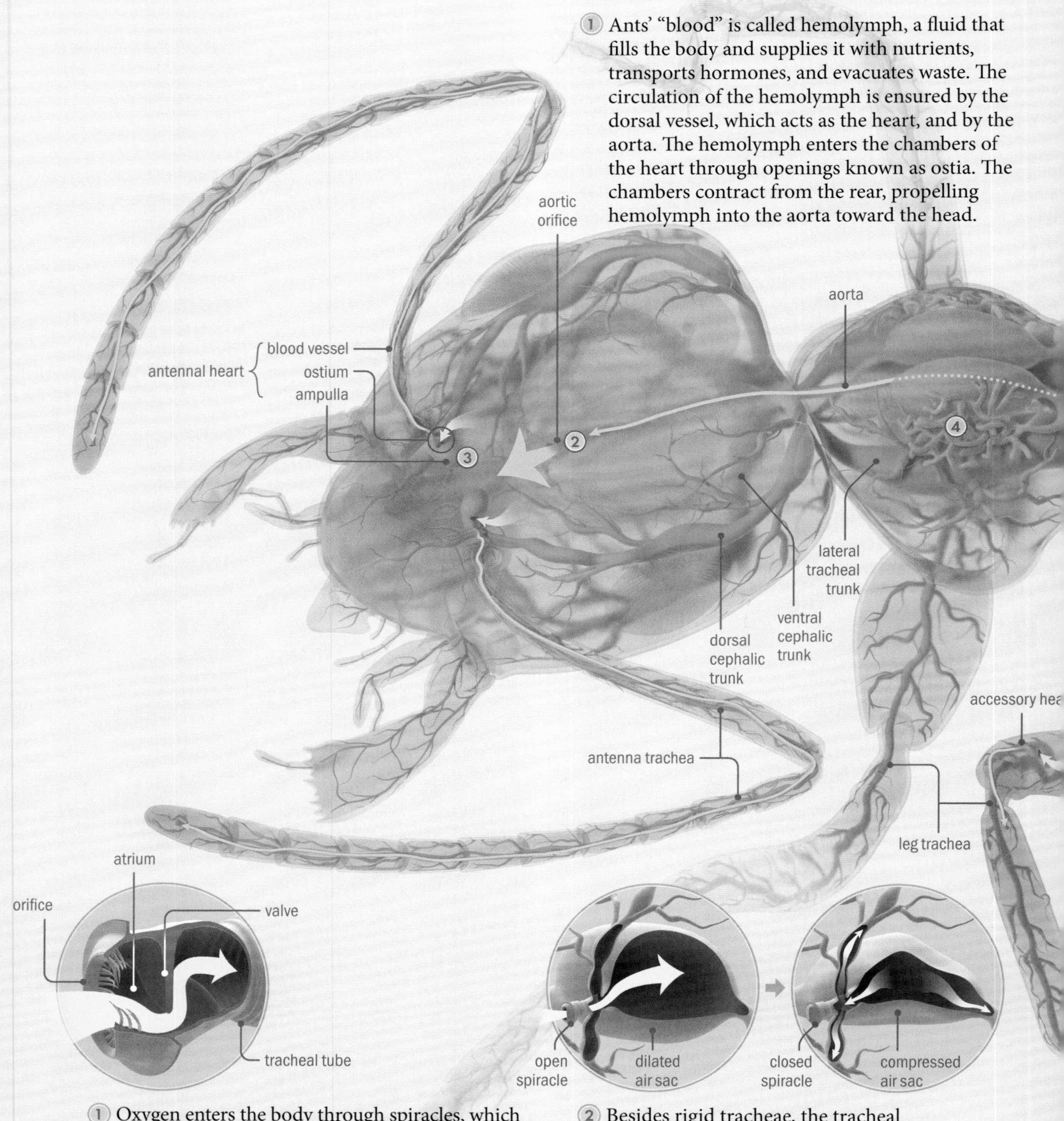

① Ants' "blood" is called hemolymph, a fluid that fills the body and supplies it with nutrients, transports hormones, and evacuates waste. The circulation of the hemolymph is ensured by the dorsal vessel, which acts as the heart, and by the aorta. The hemolymph enters the chambers of the heart through openings known as ostia. The chambers contract from the rear, propelling hemolymph into the aorta toward the head.

① Oxygen enters the body through spiracles, which are openings in the exoskeleton. At the entrance of the spiracles, hairs and a closing mechanism prevent water and foreign bodies from entering. The spiracles open into the tracheal system, a network of fine tubules that runs throughout the whole body. Oxygen flows through the system by diffusion.

② Besides rigid tracheae, the tracheal system includes soft air sacs. The air sacs are expanded or compressed by body movements, muscles, or variations in hemolymph pressure. Their rhythmic contraction promotes the transport and diffusion of oxygen throughout the body.

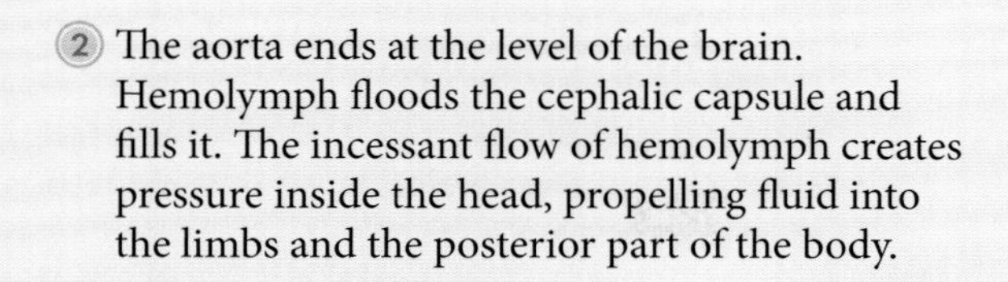

② The aorta ends at the level of the brain. Hemolymph floods the cephalic capsule and fills it. The incessant flow of hemolymph creates pressure inside the head, propelling fluid into the limbs and the posterior part of the body.

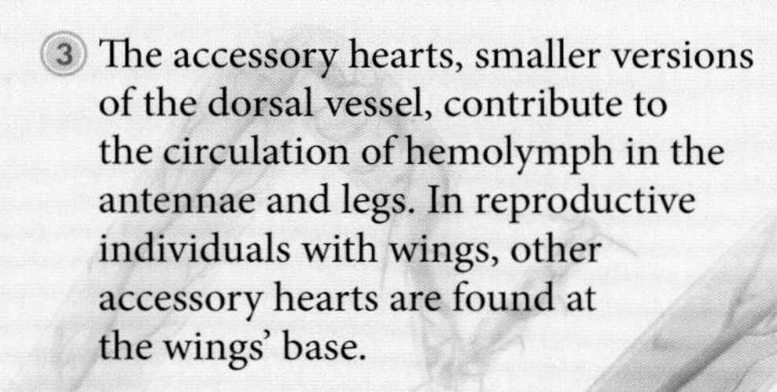

③ The accessory hearts, smaller versions of the dorsal vessel, contribute to the circulation of hemolymph in the antennae and legs. In reproductive individuals with wings, other accessory hearts are found at the wings' base.

④ The organs bathe in hemolymph. Branching extensions increase their contact surface with the liquid, where they take in nutrients and water and discharge their waste.

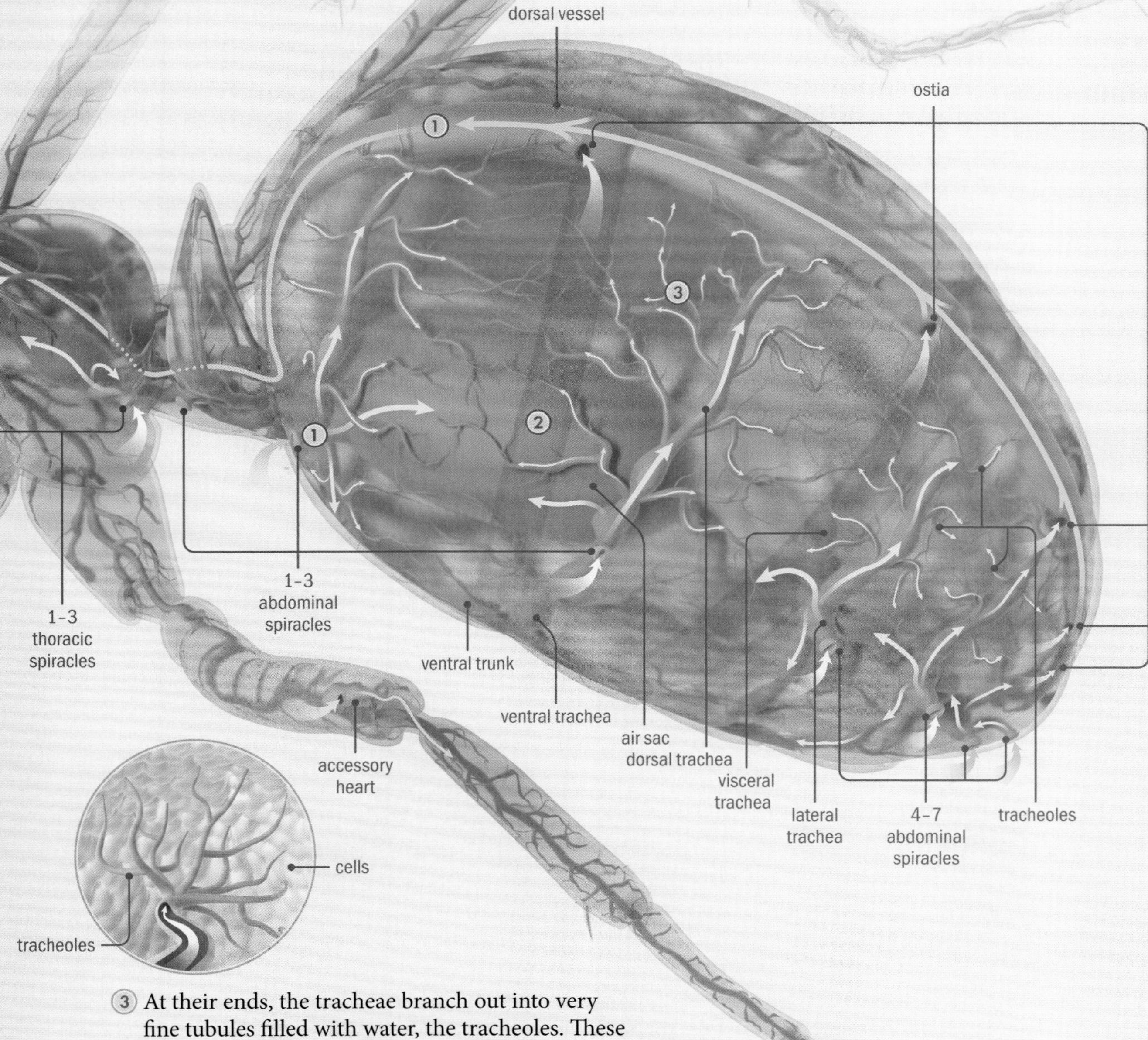

③ At their ends, the tracheae branch out into very fine tubules filled with water, the tracheoles. These end up near the cells of organs and muscles, which they supply with oxygen. The carbon dioxide produced by cellular respiration is temporarily stored in the hemolymph before being exhaled in steady puffs of water vapor from transpiration.

1.4 Nervous and digestive systems

Ants have a decentralized nervous system and a social stomach.

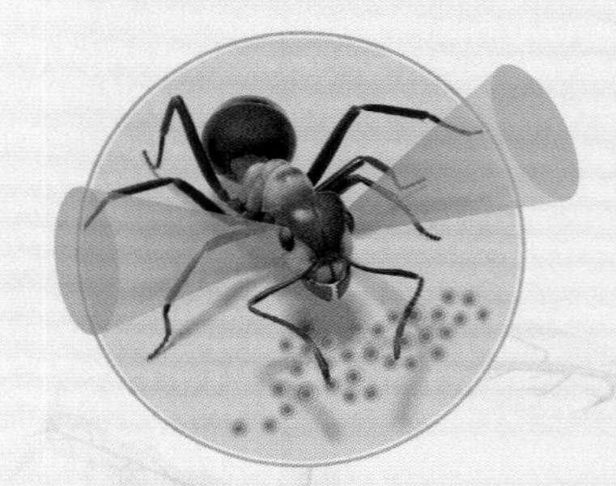

① The ants' well-developed olfactory lobes analyze information transmitted by the antennae. The optic lobes are also well developed and analyze the eyes' optical stimuli.

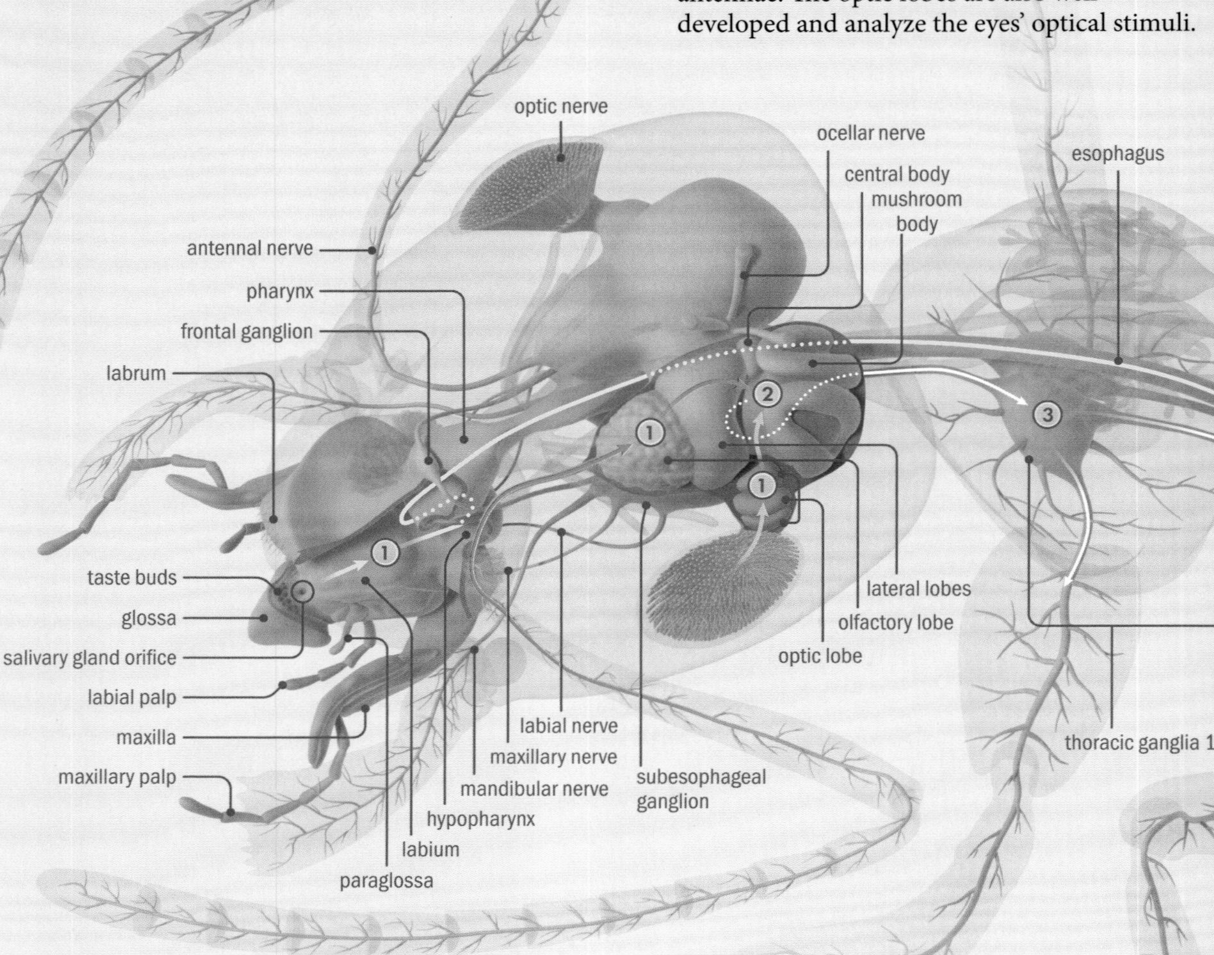

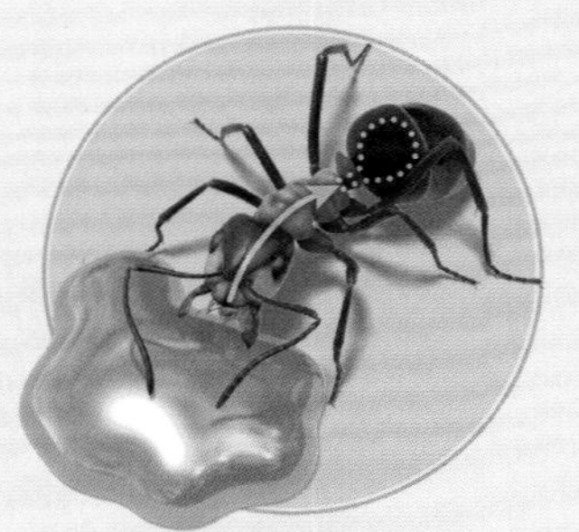

① Using their grinder-licker mouthparts, ants ingest liquid or soft food. Taste buds and the tongue's antennae-like appendages taste food before bringing it to the mouth. Before swallowing, all solid foreign bodies are filtered and collected in the infrabuccal pocket and expelled.

② Ingested food is stored in the social stomach, known as the crop. If necessary, part of the contents is regurgitated and shared with fellow creatures. The contents of a full crop can supply provisions for approximately eighty ants. These in turn share the received food until each ant is fed.

② The mushroom bodies and the central body are the headquarters of memory and learning. They constitute the associative area of the brain. Sensory information from the eyes and antennae is transmitted to the mushroom and associated bodies. This is where the ant decides how to react to sensory stimuli.

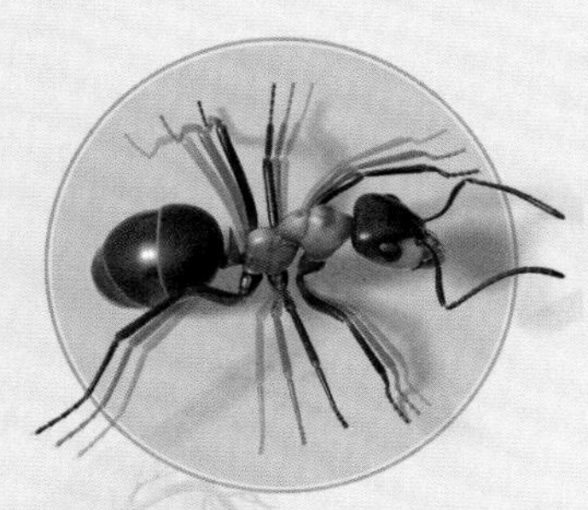

③ Beginning at the brain, the ventral chain—a chain of successive pairs of ganglia—runs through the ant's body. Each pair of ganglia controls specific parts of the body and translates orders from the brain into movement and inhibition impulses. The thoracic ganglia control the legs, while the abdominal ganglia control the digestive organs.

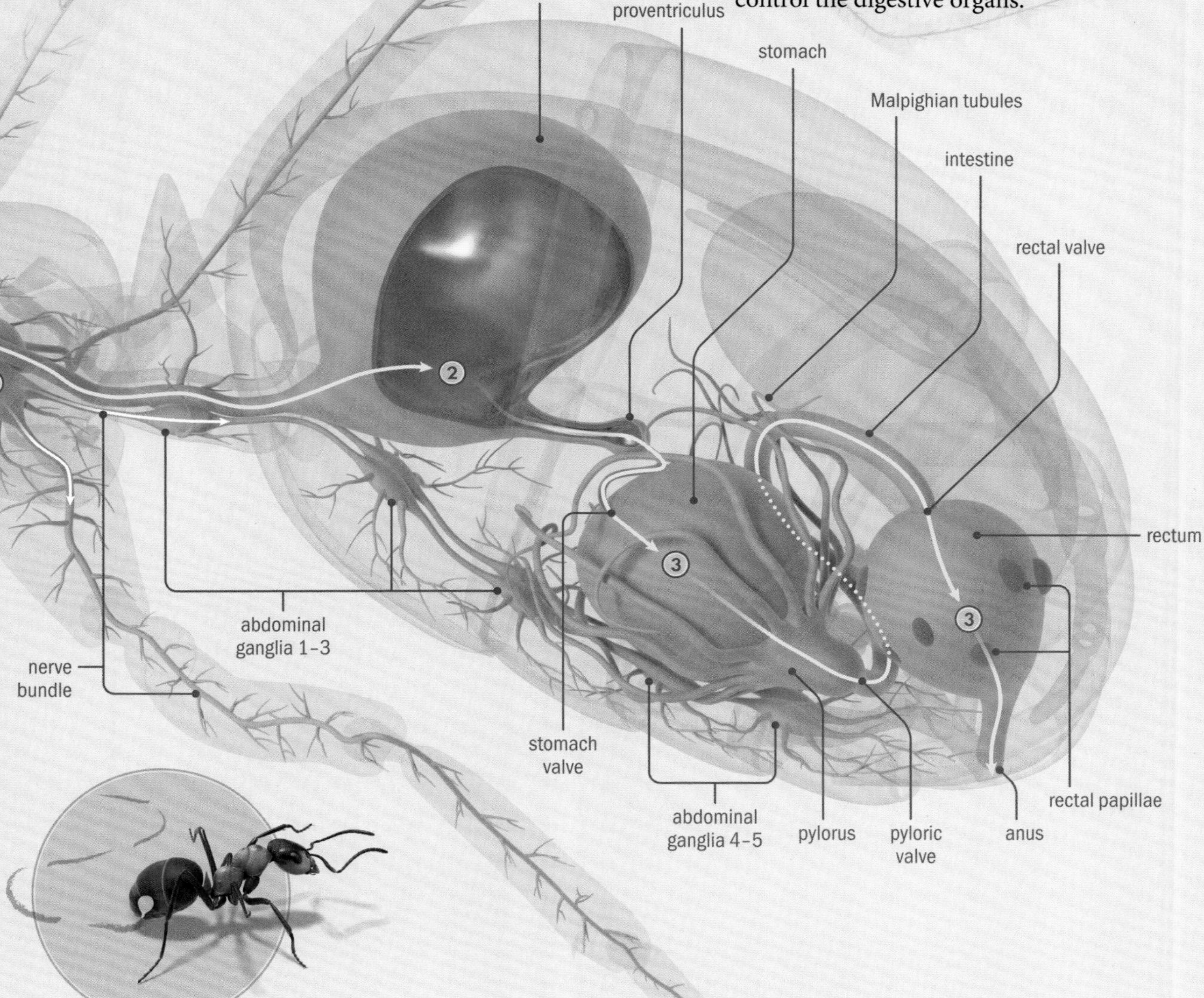

③ When the crop is nearly empty, the proventriculus pushes food for the ant's personal needs into the stomach, where the nutrients are absorbed. The undigested remains reach the pylorus, joining the toxins extracted from the blood by the Malpighian tubules. Then the waste passes into the intestine and finally into the rectum, where papillae retain the water contained in the waste. Finally, the excrement serves as an odorous substance to mark trails.

1.5 Glands

Pheromones produced by glands regulate the organism's functions and the ants' social life.

① Secretions from the prepharyngeal gland are released into the mouth. Some of them reach the crop via the esophagus. These secretions play a role in digestion or in feeding the larvae.

② Secretions from the two maxillary glands enter the mouth. The precise function of these glands is still poorly understood. They might assist in feeding the larvae.

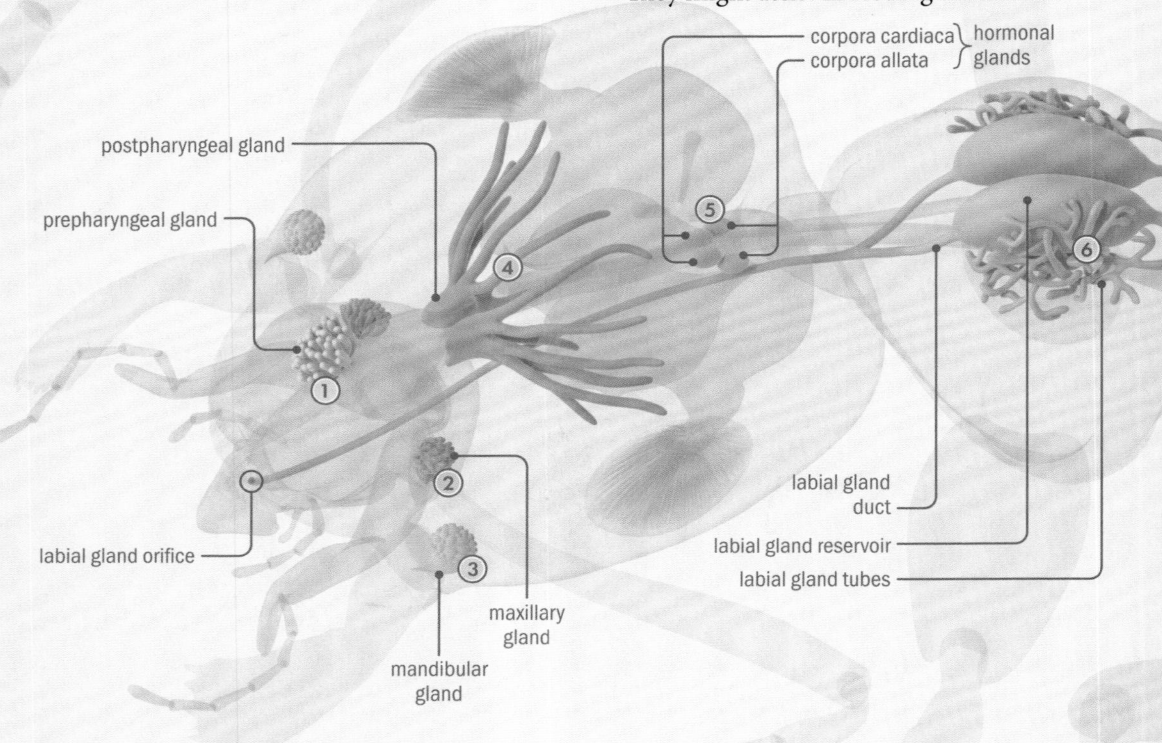

③ Secretions from the two mandibular glands reach the mouth. They may be used to coat the eggs with saliva or to lick the body.

④ The postpharyngeal gland produces saliva, which it empties into the esophagus. This gland may be involved in digestion or in producing reproductive individuals.

⑤ Hormonal gland secretions go directly into the hemolymph and spread throughout the body. Hormones act on internal organs and regulate many functions as well as metabolism.

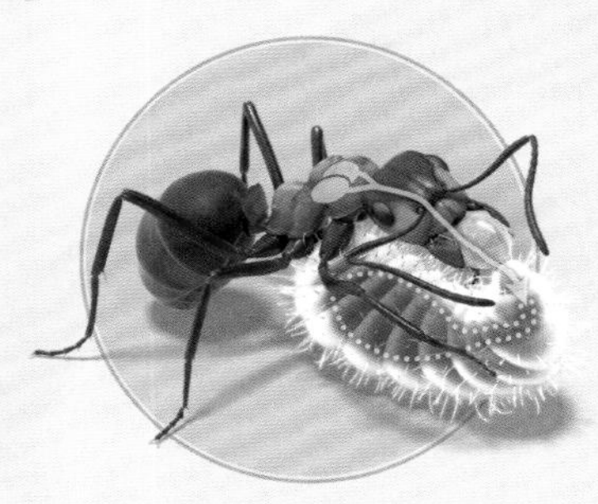

⑥ The labial gland produces a nutritive secretion that the workers give solely to the queen and the larvae that will become reproductive individuals. The amount of secretion given regulates the queen's fertility. This secretion contributes to the development of “female” larvae into queens.

⑦ The metathoracic gland emits its secretion outside the organism, where the legs apply it to the surface of the body. In workers, it has an antiseptic effect and protects against infections. In the queen, it produces the royal scent. By taking care of her, the workers receive the scent and transmit it by bodily contact to all the inhabitants of the anthill. Ants can thus recognize each other as members of the same colony.

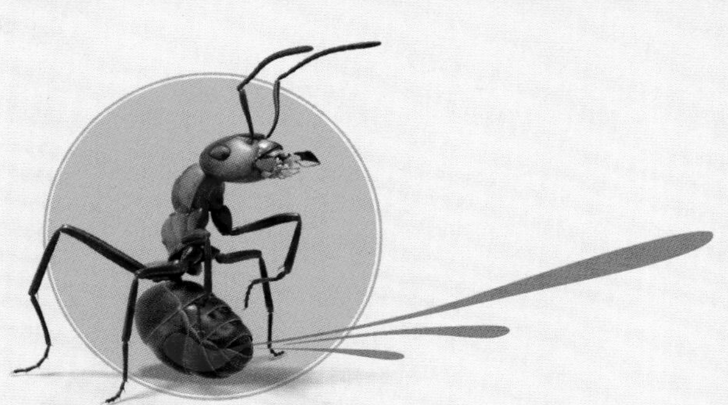

⑧ The poison gland's painful secretion (formic acid) is a biological defense weapon. The worker projects it directly or from a distance into an opponent's wounds. Penetrating into the body of the victim, it destroys the tissues and has a paralyzing effect. Its odor also acts as an alarm: it excites the other workers, who join in the fight.

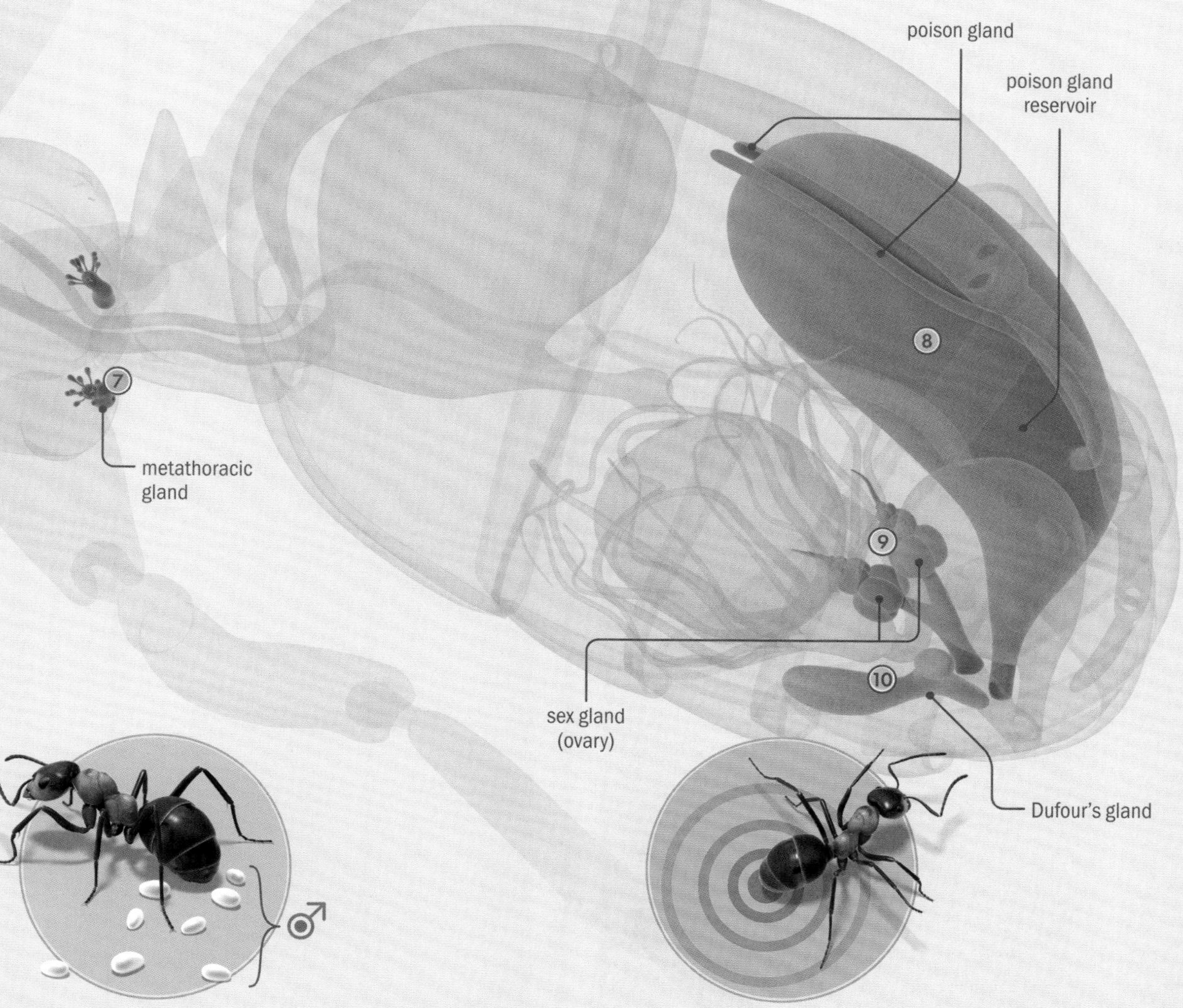

⑨ The ovaries produce eggs in reproductive individuals. The workers' ovaries are much smaller and are functional only in young workers. When the queen dies, the young workers can lay eggs. As they do not mate, the eggs are not fertilized and only produce males.

⑩ The secretion of the Dufour's gland informs the other workers of the presence of food and encourages them to leave the nest in search of it. If the secretion is sprayed together with formic acid, it increases its alarm effect. In queens, it serves as a sexual scent to attract males during the nuptial flight.

1.6 Sex and caste anatomy

The ants' external and internal structure reflects the functions of the sexes and the different castes. The lives of workers comprise two phases characterized by different activities, to which the functions of their organs adapt. For their part, males and female queens are exclusively adapted to reproduction.

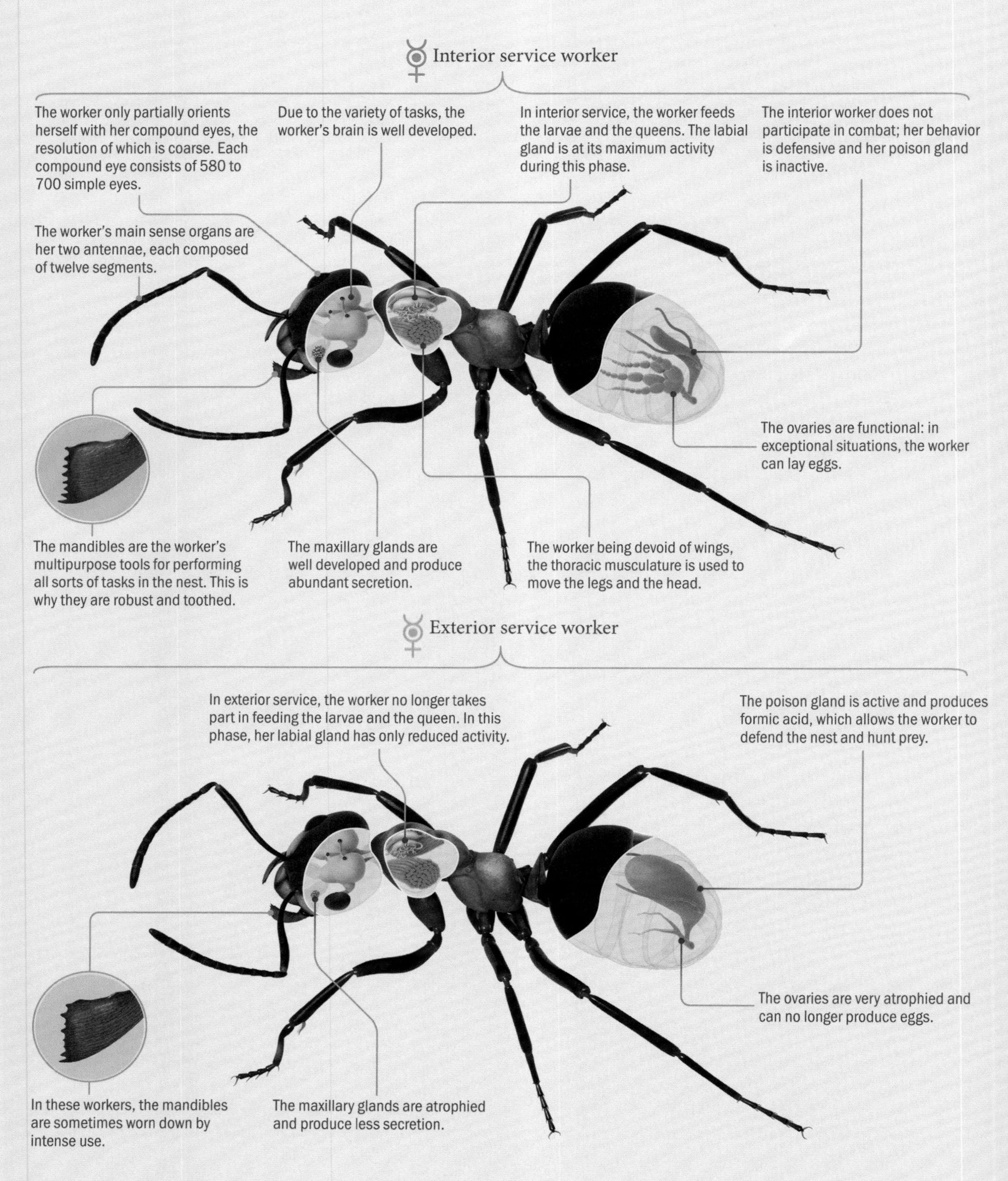

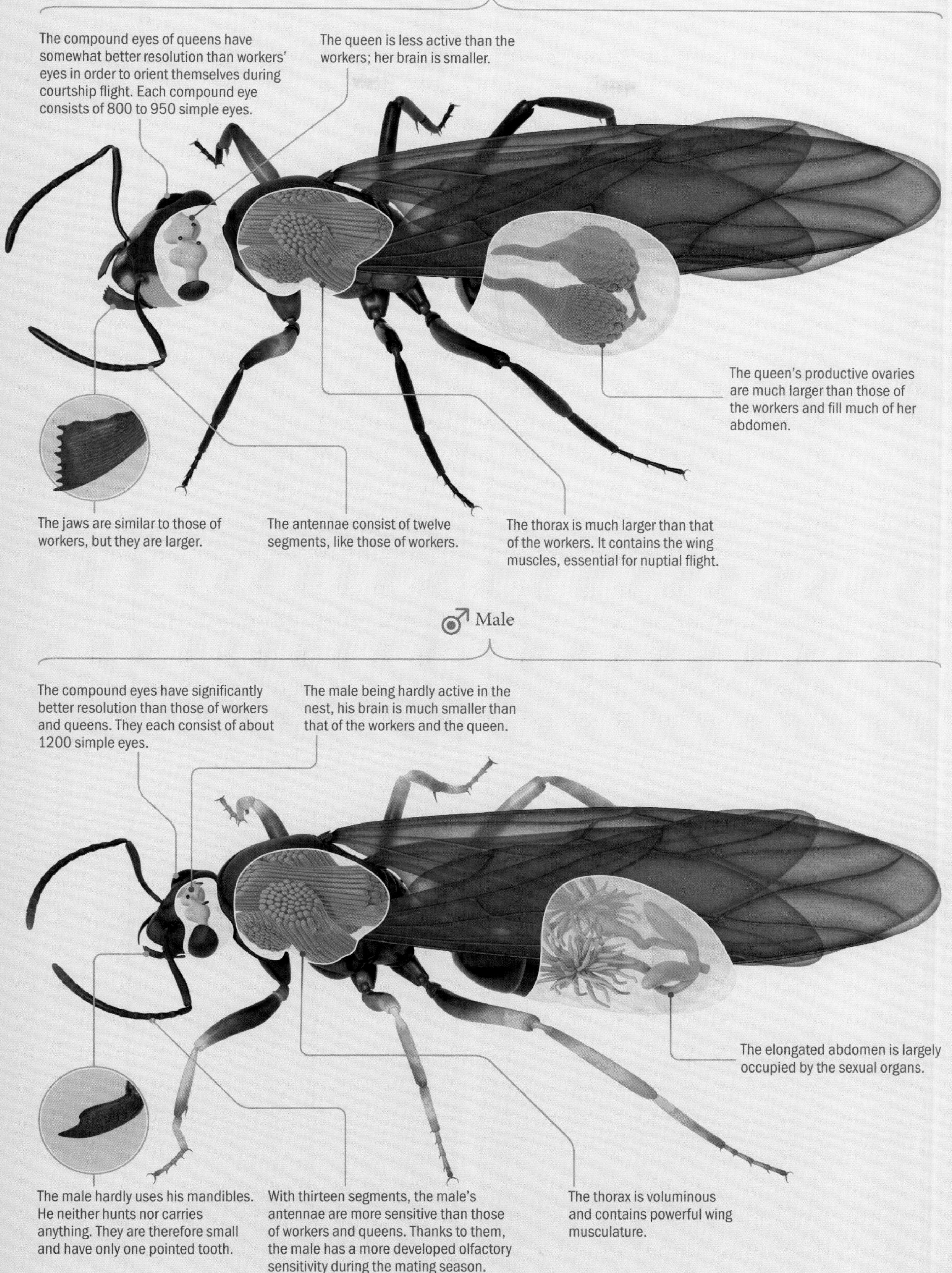
Queen
The compound eyes of queens have somewhat better resolution than workers' eyes in order to orient themselves during courtship flight. Each compound eye consists of 800 to 950 simple eyes.
The queen is less active than the workers; her brain is smaller.
The queen's productive ovaries are much larger than those of the workers and fill much of her abdomen.
The jaws are similar to those of workers, but they are larger.
The antennae consist of twelve segments, like those of workers.
The thorax is much larger than that of the workers. It contains the wing muscles, essential for nuptial flight.
Male
The compound eyes have significantly better resolution than those of workers and queens. They each consist of about 1200 simple eyes.
The male being hardly active in the nest, his brain is much smaller than that of the workers and the queen.
The elongated abdomen is largely occupied by the sexual organs.
The male hardly uses his mandibles. He neither hunts nor carries anything. They are therefore small and have only one pointed tooth.
With thirteen segments, the male's antennae are more sensitive than those of workers and queens. Thanks to them, the male has a more developed olfactory sensitivity during the mating season.
The thorax is voluminous and contains powerful wing musculature.

2.1 Mating season

In May and June, the red wood ants take off en masse to mate. During this period, especially in sunny and muggy weather, males and females from the many colonies gather in their nests, waiting for the flight signal.

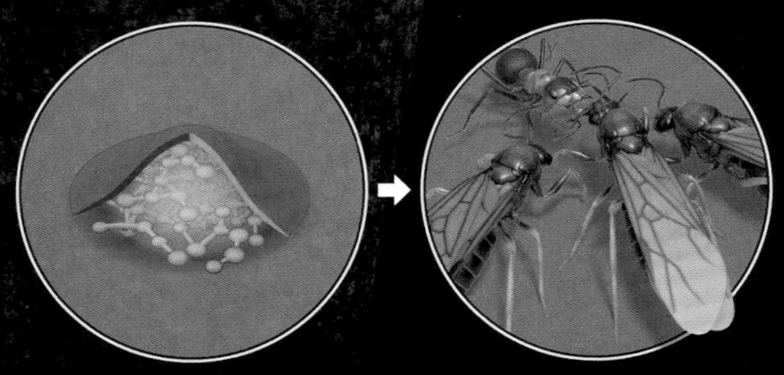

Males and females develop separately. Males are mainly born in small, still undeveloped nests that produce little internal heat. When mating season arrives, they leave the interior of the anthill. Before flying away, they are fed one last time by the workers.

The workers restrain the males by the wings in order to prevent their premature flight.

An atmospheric depression at the end of the afternoon gives the signal for takeoff. Swarms of males leave the nest. Their two mandibular glands secrete pheromones that diffuse into the air, informing ants in nearby nests of their flight.
Males that try to leave the nest too soon are caught by the workers and brought back to the nest.

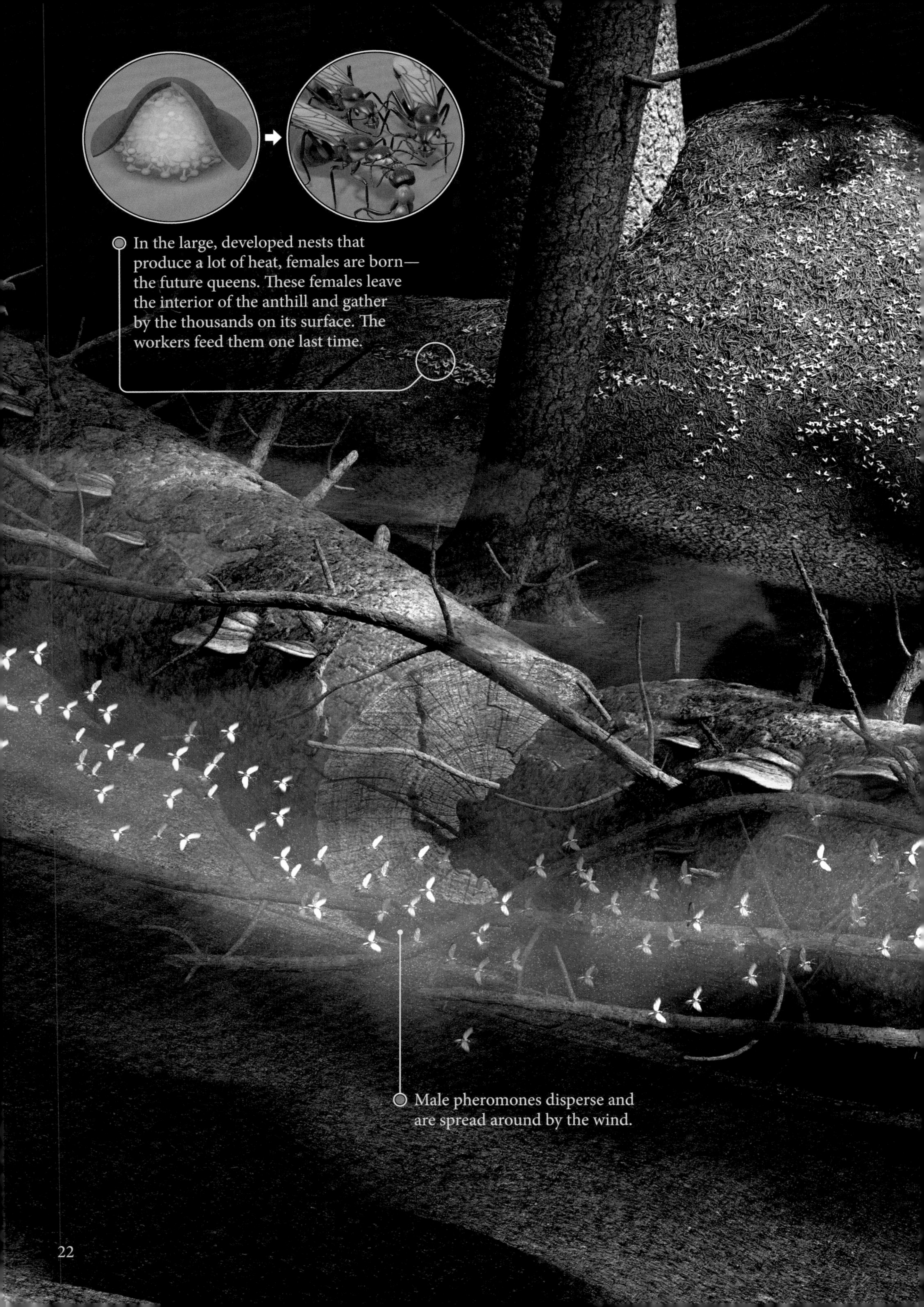
In the large, developed nests that produce a lot of heat, females are born—the future queens. These females leave the interior of the anthill and gather by the thousands on its surface. The workers feed them one last time.
Male pheromones disperse and are spread around by the wind.

As soon as the females' antennae perceive the males' pheromones, they become agitated and prepare to leave the nest.
If a female attempts to leave the nest too early, she is caught by a worker and carried back to the nest.

2.2 Nuptial flight

The meeting point between the two swarms is a small clearing. This is where they come together in a single swarm for the nuptial flight.

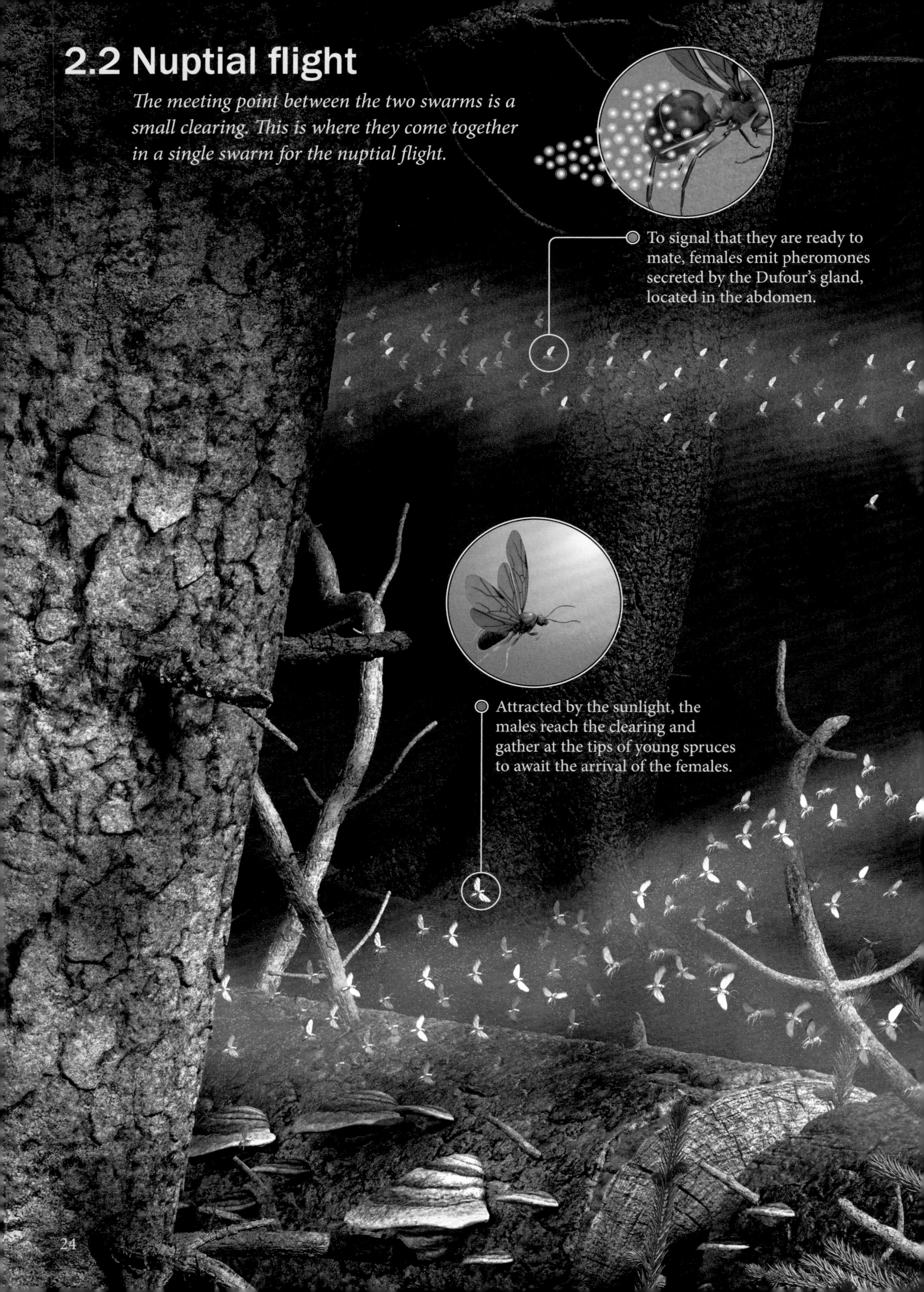

In turn, the females' pheromones spread to the surrounding area.
As soon as the males sense the females' pheromones, they take flight and form a large swarm with them.

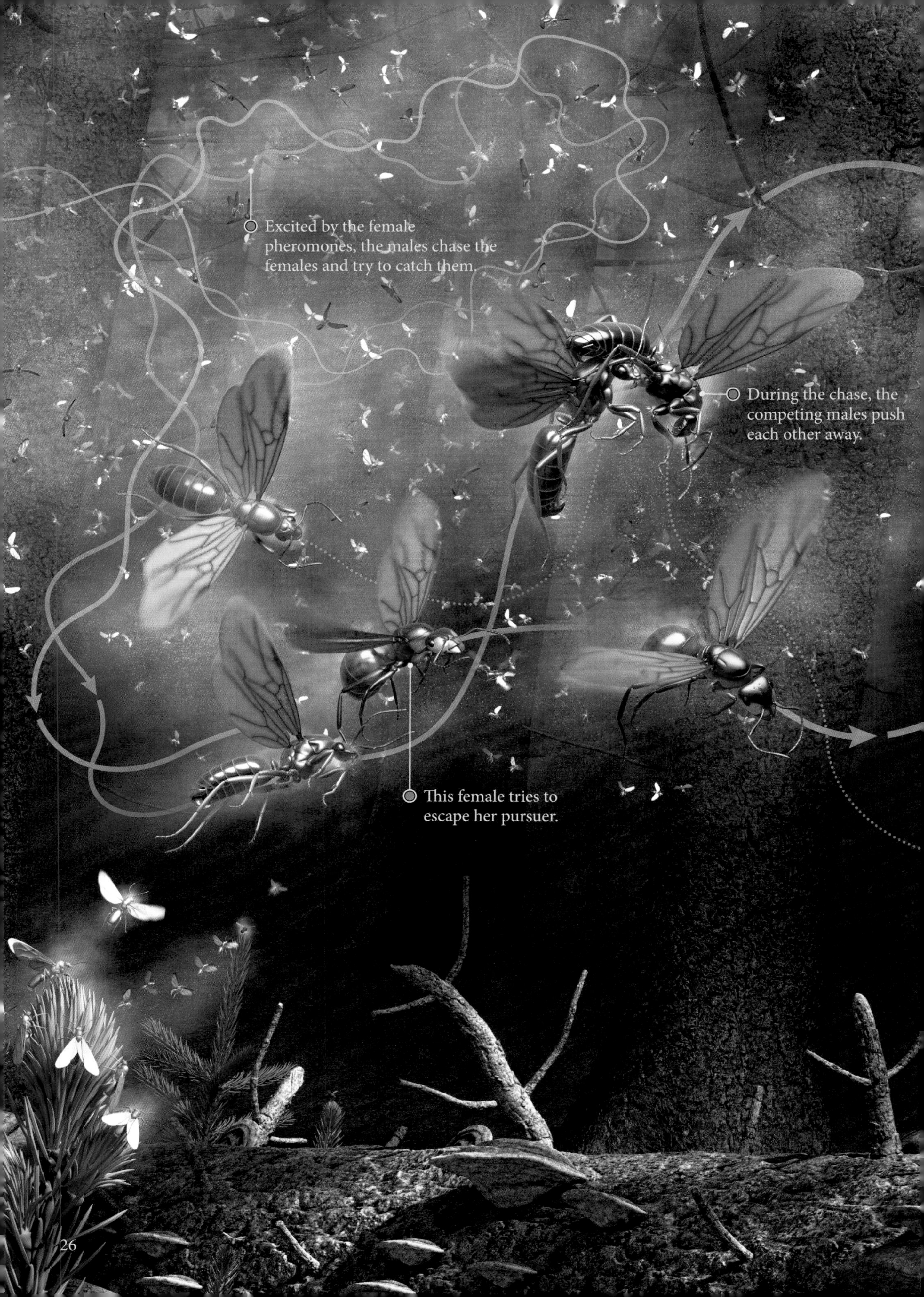
Excited by the female pheromones, the males chase the females and try to catch them.
During the chase, the competing males push each other away.
This female tries to escape her pursuer.

This male positions himself above the coveted female.
He quickly approaches and grabs her back with his claws.
The female cannot support the male's weight and they fall to the ground.

2.3 Mating and search for a nest

During mating, the female receives a supply of sperm from the males. As soon as she has established a nest, she will use it to produce the anthill's inhabitants.

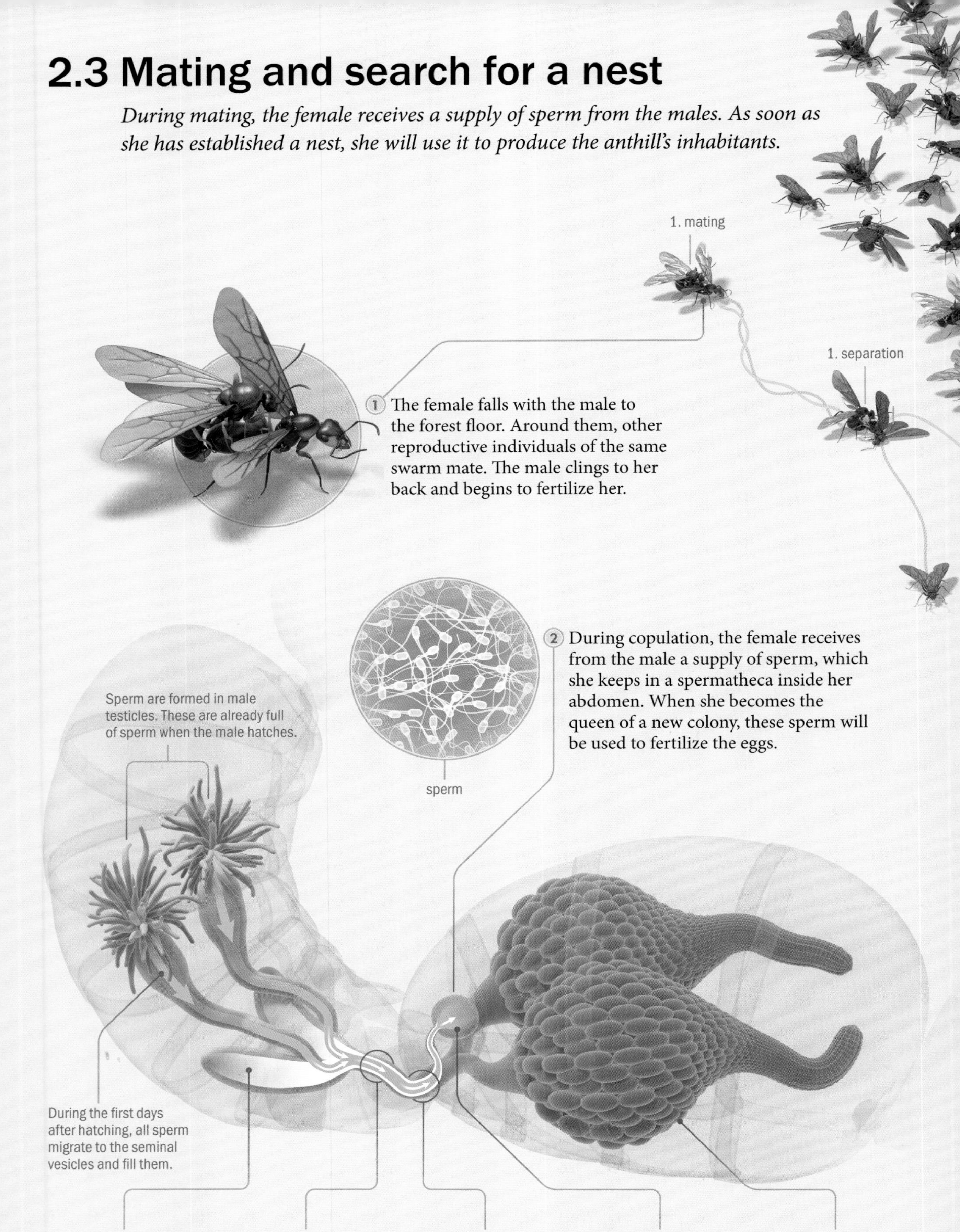

① The female falls with the male to the forest floor. Around them, other reproductive individuals of the same swarm mate. The male clings to her back and begins to fertilize her.

② During copulation, the female receives from the male a supply of sperm, which she keeps in a spermatheca inside her abdomen. When she becomes the queen of a new colony, these sperm will be used to fertilize the eggs.

At the same time, the two accessory glands develop and produce seminal fluid or semen.

During mating, sperm and semen mix and are projected out through the seminal canal and penis.

The female's copulatory orifice accommodates the male's penis and the semen enters her body.

The sperm enter the spermatheca. Glands there produce a secretion that nourishes them and keeps them alive.

The eggs develop in the female's ovaries. When they are laid, they are selectively fertilized by sperm from the spermatheca.

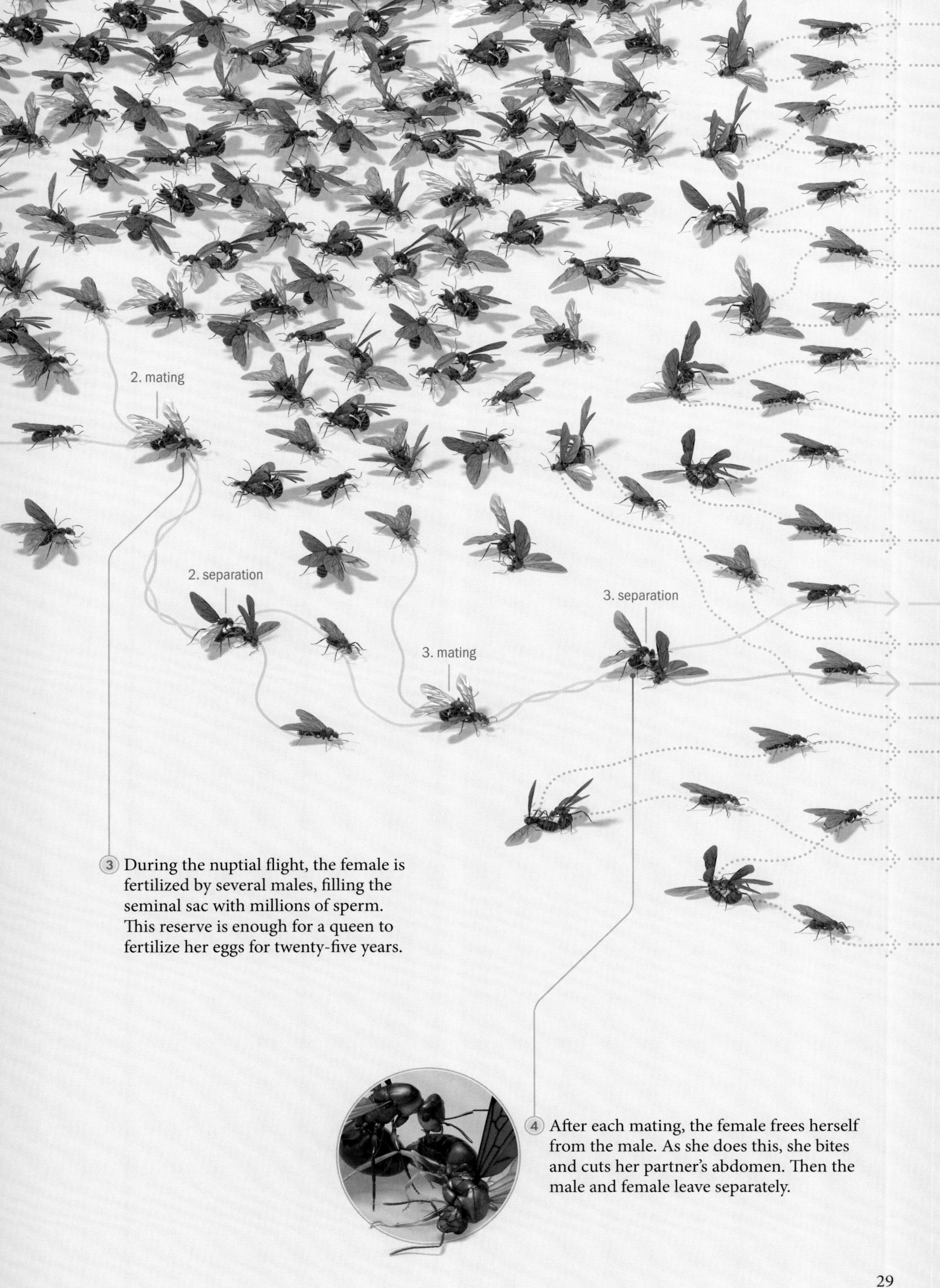

③ During the nuptial flight, the female is fertilized by several males, filling the seminal sac with millions of sperm. This reserve is enough for a queen to fertilize her eggs for twenty-five years.

④ After each mating, the female frees herself from the male. As she does this, she bites and cuts her partner's abdomen. Then the male and female leave separately.

6 Having mated, the males have fulfilled their biological function. They do not return to the nest and behave with indifference. Soon after, they die or are eaten by animals.
5 After the last mating, the young fertilized queens land en masse on the ground. The long and dangerous search for a place for the nest begins. Few queens survive this exploration.
7 The young fertilized queen is excluded from returning to her natal nest. Indeed, colonies of red wood ants do not accept competitors to their queens. Anyone who returns would be considered an enemy and attacked by the workers. The young queen must therefore find her own nest.

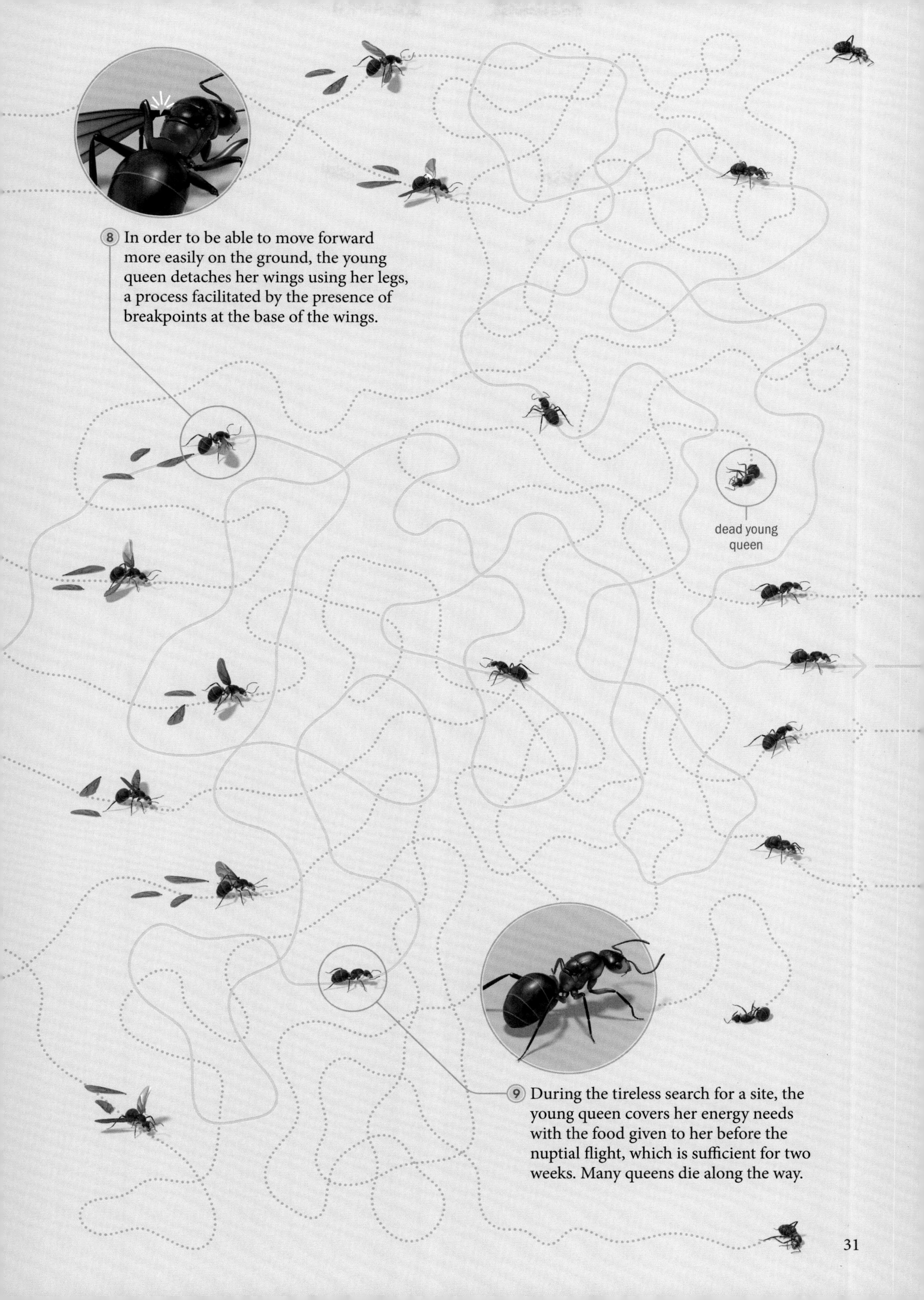

8 In order to be able to move forward more easily on the ground, the young queen detaches her wings using her legs, a process facilitated by the presence of breakpoints at the base of the wings.

9 During the tireless search for a site, the young queen covers her energy needs with the food given to her before the nuptial flight, which is sufficient for two weeks. Many queens die along the way.

2.4 Nest establishment

*During its evolution, the red wood ant (*Formica rufa*) lost the ability to build a nest. For this undertaking, the young queen needs help from host ants. She must therefore find a host whose nest she can colonize.*

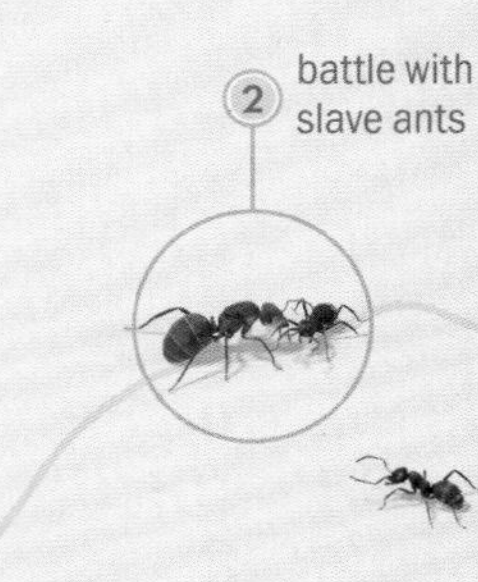

1 The young queen discovers a nest inhabited by dark gray slave ants (*Formica fusca*). Like the red wood ant, this species likes to build its nests under the trunks of trees. It's an ideal host for the young queen.

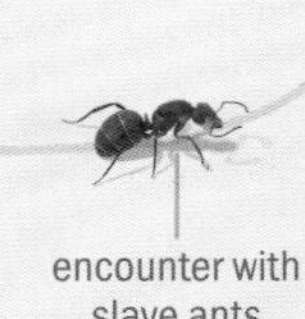

2 For the slave workers, the approaching young queen is an enemy. If there are many workers and a fight ensues, she won't survive. But if the queen can elude them, she can overcome some workers thanks to her larger size.

3 As soon as the young queen finds an entrance, she enters the nest and looks for the chamber in which the slave ants' queen is held.

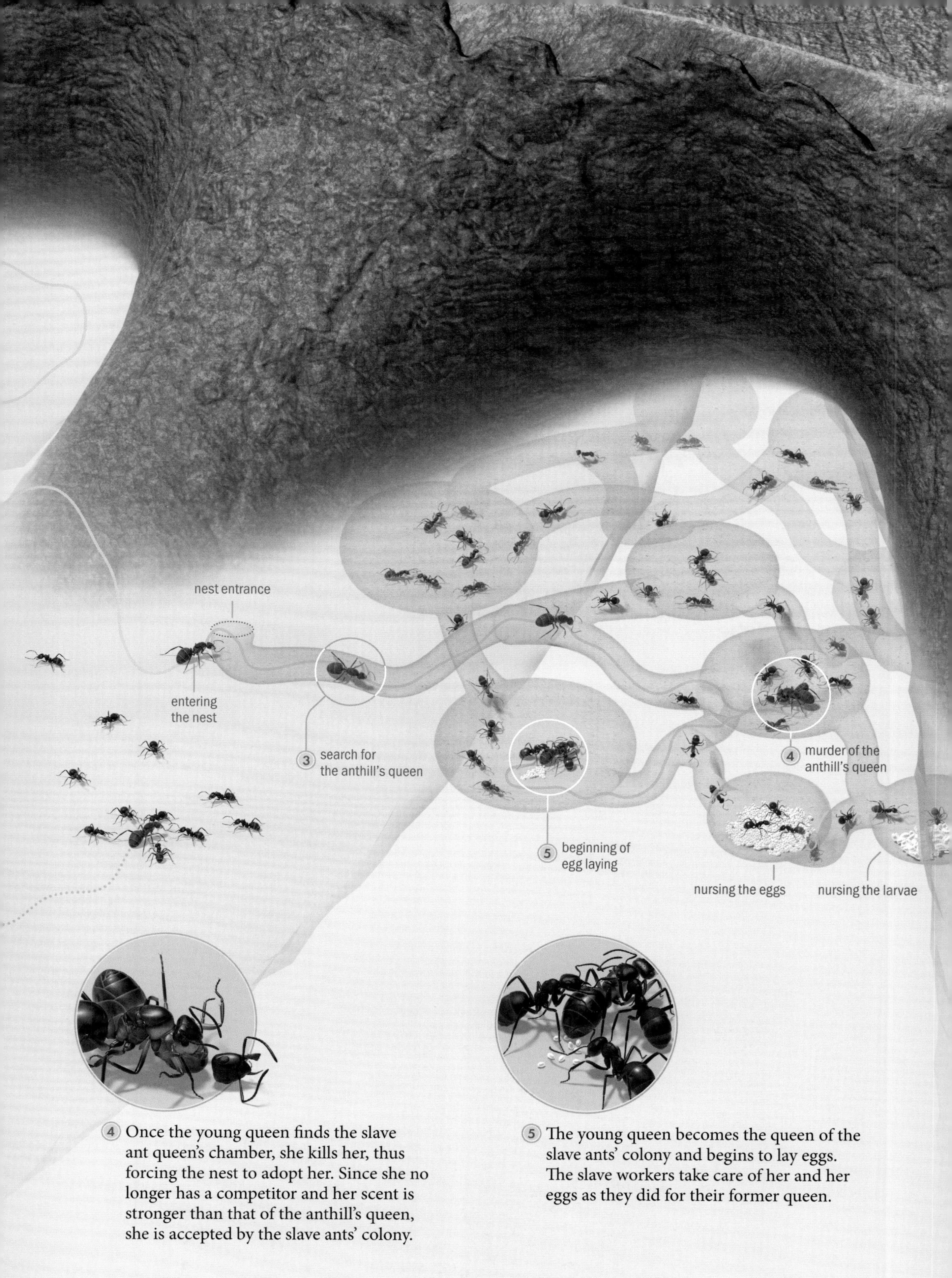

4 Once the young queen finds the slave ant queen's chamber, she kills her, thus forcing the nest to adopt her. Since she no longer has a competitor and her scent is stronger than that of the anthill's queen, she is accepted by the slave ants' colony.

5 The young queen becomes the queen of the slave ants' colony and begins to lay eggs. The slave workers take care of her and her eggs as they did for their former queen.

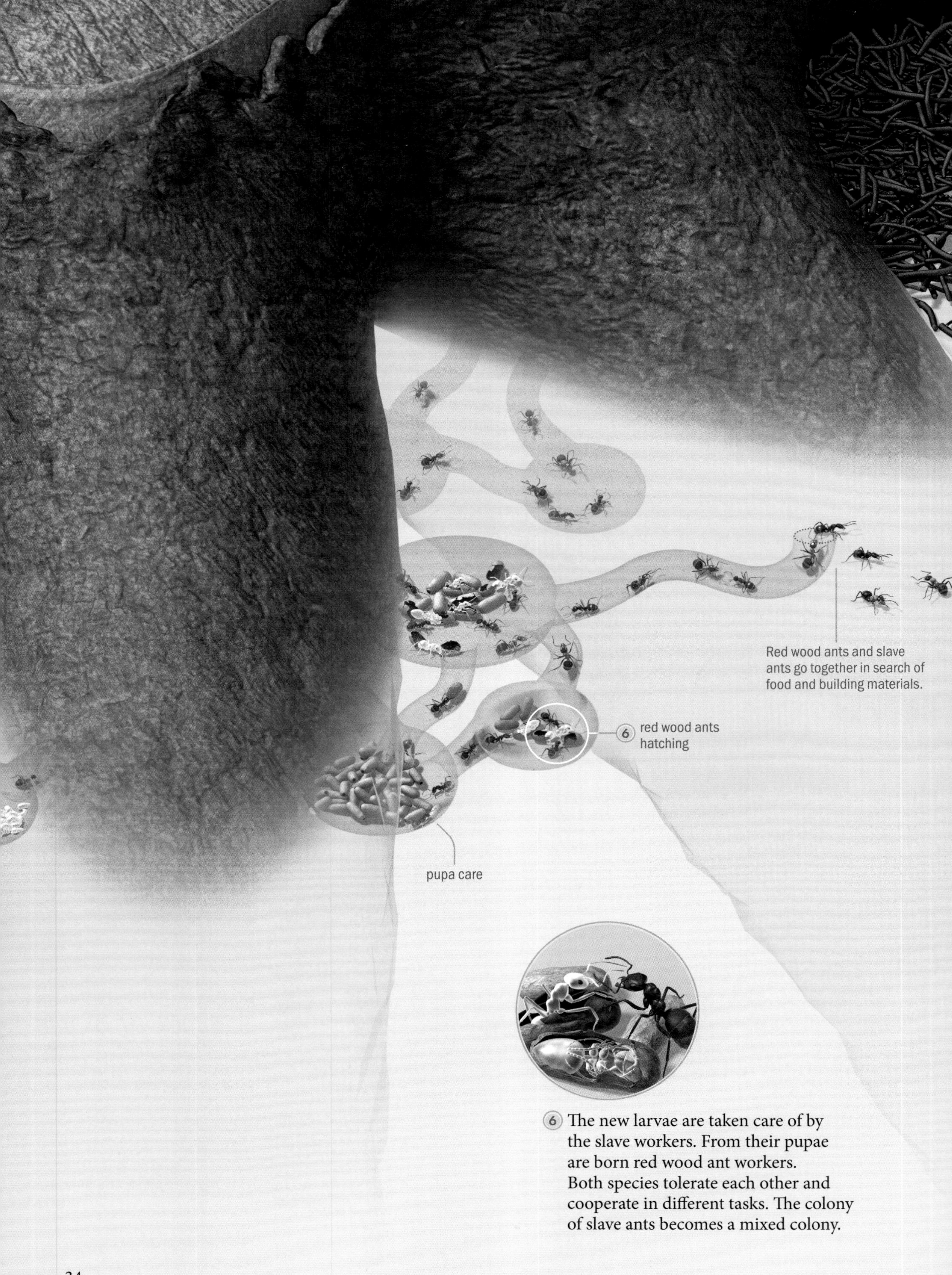

⑥ The new larvae are taken care of by the slave workers. From their pupae are born red wood ant workers. Both species tolerate each other and cooperate in different tasks. The colony of slave ants becomes a mixed colony.

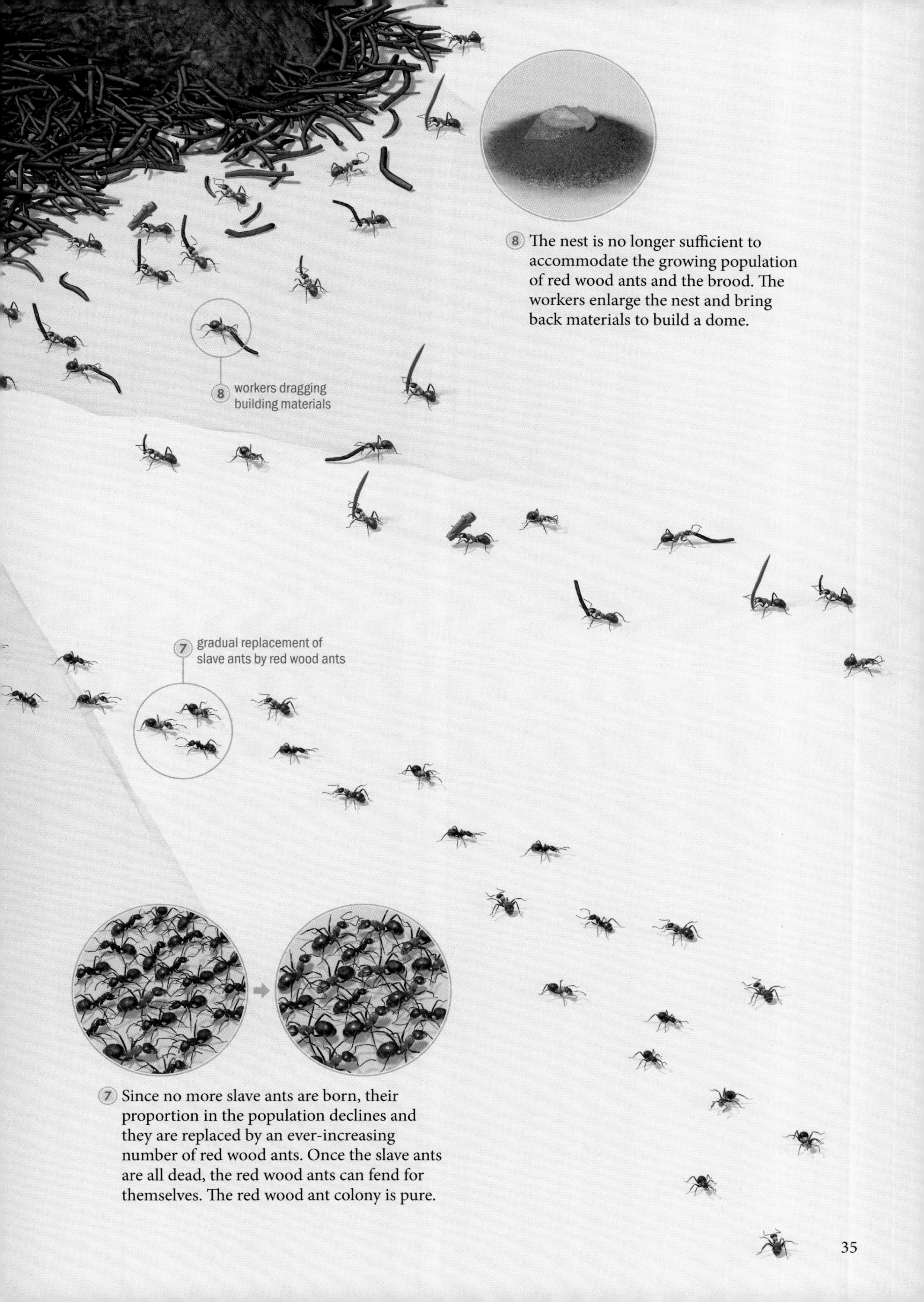

8 The nest is no longer sufficient to accommodate the growing population of red wood ants and the brood. The workers enlarge the nest and bring back materials to build a dome.

7 Since no more slave ants are born, their proportion in the population declines and they are replaced by an ever-increasing number of red wood ants. Once the slave ants are all dead, the red wood ants can fend for themselves. The red wood ant colony is pure.

3.1 Construction materials

One of the workers' main tasks is the construction and maintenance of the anthill dome, whose materials come from the surrounding forest.

① The worker's main tools are their serrated upper mandibles. Thanks to the power of the mandibular muscles and their expansive opening, the workers can grasp twigs and pull them to the nest. They work together to carry the biggest ones.

4 Near the nest, materials are passed to other workers, who carry them to their final location in the dome.
3 Building materials are collected near the anthill. Transporting large items can take several days.
twig
solidified resin
center of gravity
2 The worker's center of gravity is in her posterior half, which allows her to carry heavy loads. Moreover, her six legs give her great stability. She can hold bulky loads at an angle above the body so that the load's center of gravity is close to the body's center of gravity.

5 The materials are brought in and deposited on the dome's surface. They are used to repair damage and maintain an even surface.
6 The outer cover of the dome is a layer of thin materials about 10 cm thick. Densely packed, these materials are held together by tree resin, which softens and sticks in the sun. This cover protects the inside of the nest from rain and wind and serves as insulation. Its maintenance is of vital importance, because only an intact cover can keep the heat inside the nest.
spruce needle
bud scale
pebble
plant debris

3.2 Nest architecture

The different stages of brood development (eggs, larvae, and pupae) require different temperatures and humidity levels. The workers are constantly transporting the brood to rooms where the conditions are the most favorable.

The interior of the dome is made of coarse materials and surrounds a central tree stump. A dense network of rooms serves as chambers, incubators, and storage areas.

The central stump gives stability to the dome. The rooms allow the queen to be sheltered when the dome is destroyed by animals in search of food, for example, wild boars. The precious eggs are well sheltered under the stump.

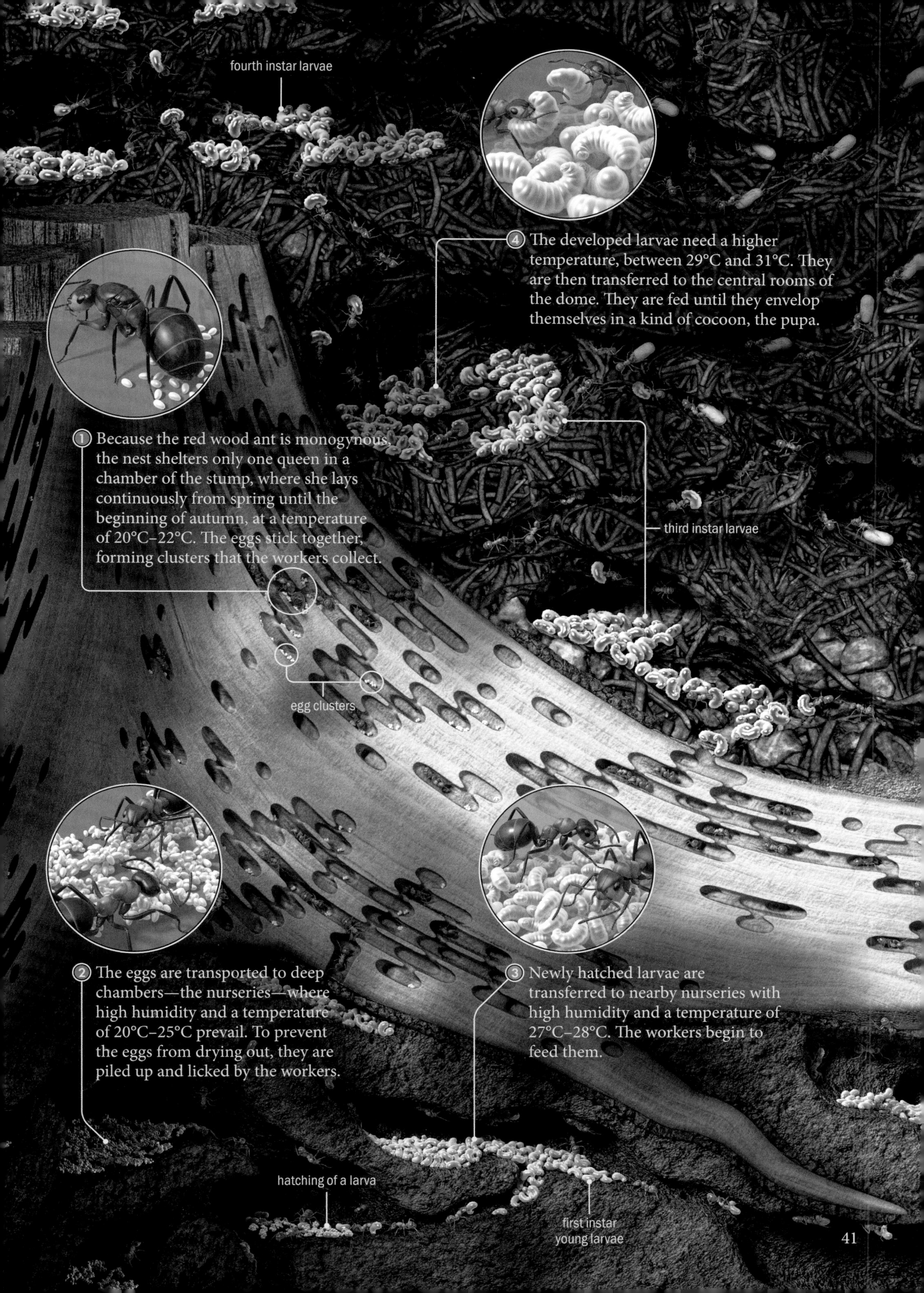
fourth instar larvae
4 The developed larvae need a higher temperature, between 29°C and 31°C. They are then transferred to the central rooms of the dome. They are fed until they envelop themselves in a kind of cocoon, the pupa.
1 Because the red wood ant is monogynous, the nest shelters only one queen in a chamber of the stump, where she lays continuously from spring until the beginning of autumn, at a temperature of 20°C–22°C. The eggs stick together, forming clusters that the workers collect.
third instar larvae
egg clusters
2 The eggs are transported to deep chambers—the nurseries—where high humidity and a temperature of 20°C–25°C prevail. To prevent the eggs from drying out, they are piled up and licked by the workers.
3 Newly hatched larvae are transferred to nearby nurseries with high humidity and a temperature of 27°C–28°C. The workers begin to feed them.
hatching of a larva
first instar young larvae

5 The pupae in their cocoons require a dry and warm environment. They are brought up to the outer areas of the anthill, at a temperature of 29°C–31.5°C. If prolonged rains moisten and cool these areas, the cocoons are briefly brought to the surface to dry in the sun.
workers hatching
Recently hatched ants have a soft, clear cuticle. This hardens after a few days and takes on its characteristic color.
Young pupae have a pale cocoon.
Older pupae have a dark cocoon.
Due to the dome's constant movement, coarse materials such as pebbles and large twigs end up piling up at the bottom of the nest.
second instar larvae
The subterranean part of the nest extends far into the ground, up to two meters deep and wide in large nests. Underground, the temperature is quite low and relatively constant. If it's too hot or too cold on the surface, the workers retreat to these subterranean areas.
second instar larvae

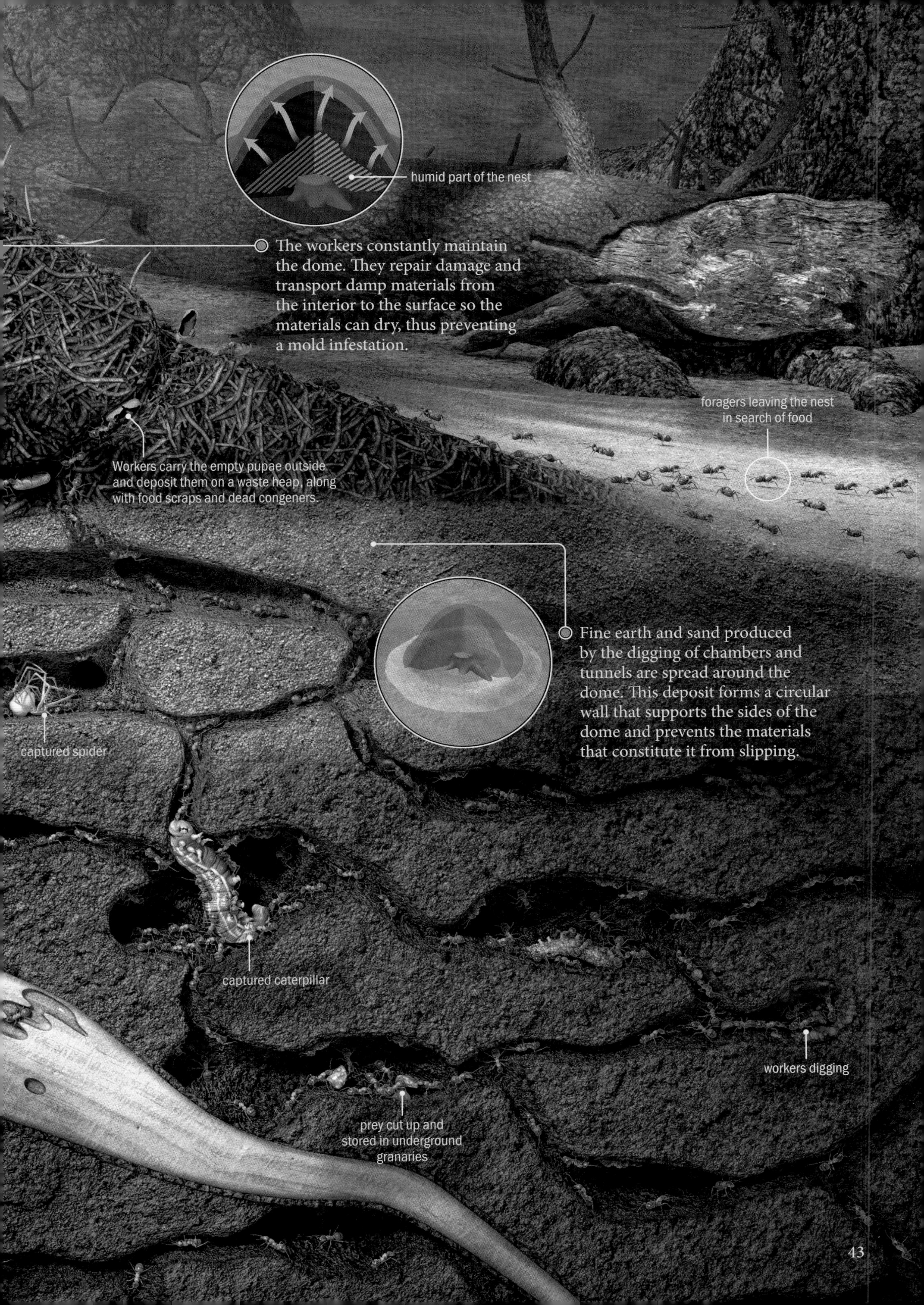

The workers constantly maintain the dome. They repair damage and transport damp materials from the interior to the surface so the materials can dry, thus preventing a mold infestation.

Fine earth and sand produced by the digging of chambers and tunnels are spread around the dome. This deposit forms a circular wall that supports the sides of the dome and prevents the materials that constitute it from slipping.

3.3 Annual cycle | Spring

A red wood ant nest is a dynamic structure that the workers are constantly adapting to temperature and weather fluctuations. In the spring, the colony awakens from its winter rest. It begins by restoring the nest to working order and warming it up.

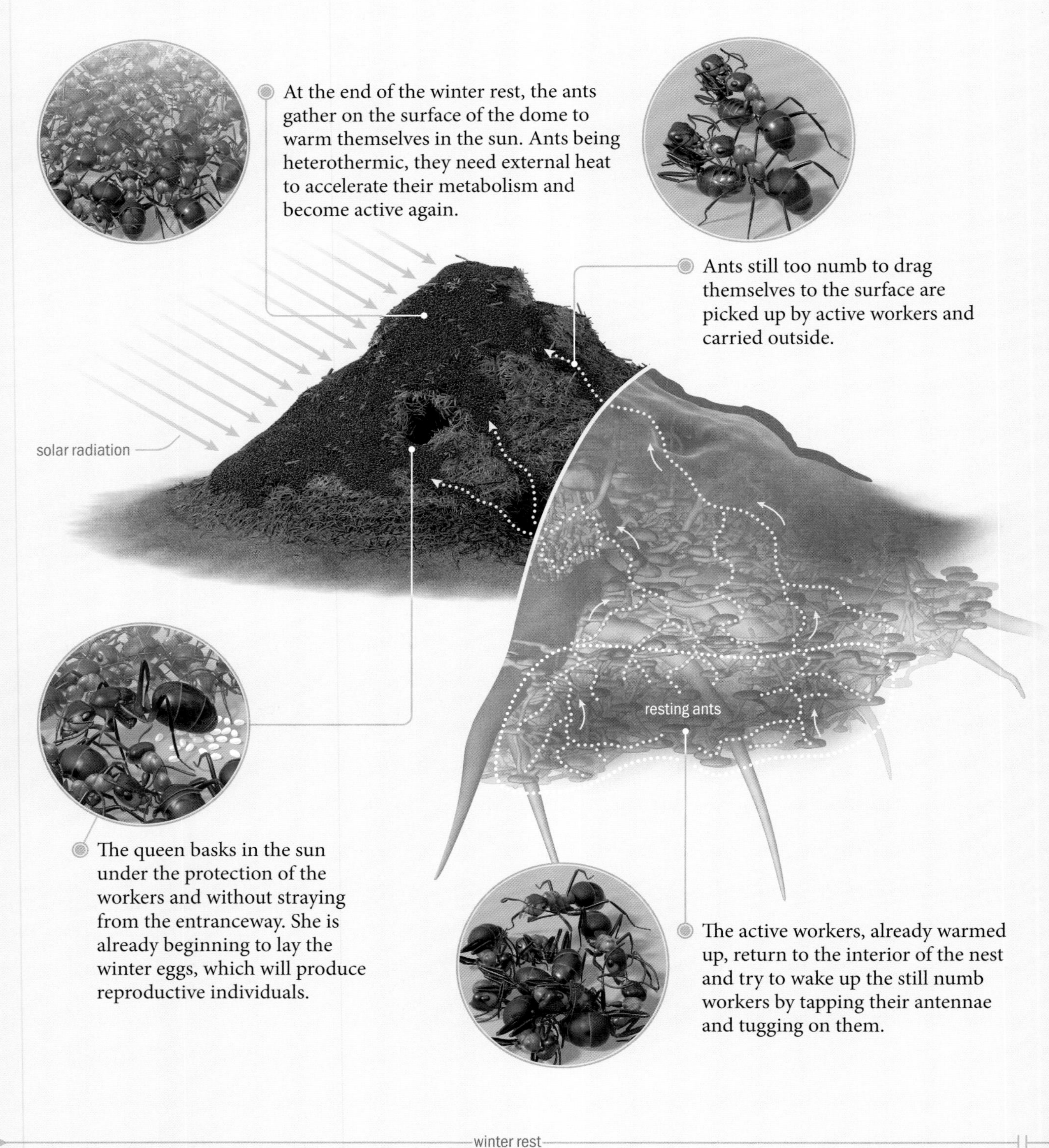

After this first sunbathing, some of the workers begin to rebuild the dome, clearing away the dirty layer covering it. Thus uncovered, the inside of the nest can dry out in the sun. The interior materials are spread around to dry too.

The workers who are not busy rebuilding the nest gather in small clusters on the surface to warm themselves in the sun. Once their body temperature reaches 30°C–34°C, they immediately rush inside the nest.

workers leaving the nest in all directions in search of food

evaporating water

inward heat transfer

return to surface

Warmed-up workers transmit their gathered heat inside the nest. Like hot water bottles, they warm the anthill. They then return to the surface to warm up, and the heat transfer begins again.

The first laid eggs hatch in the spring and generally give rise to reproductive individuals, who are cared for by the workers. In late spring and early summer, they leave the nest to mate.

spring sun bath

raising reproductive individuals

April

May

3.4 Summer

The anthill reaches a stage of intense activity and growth. Workers adapt the nest to the summer heat.

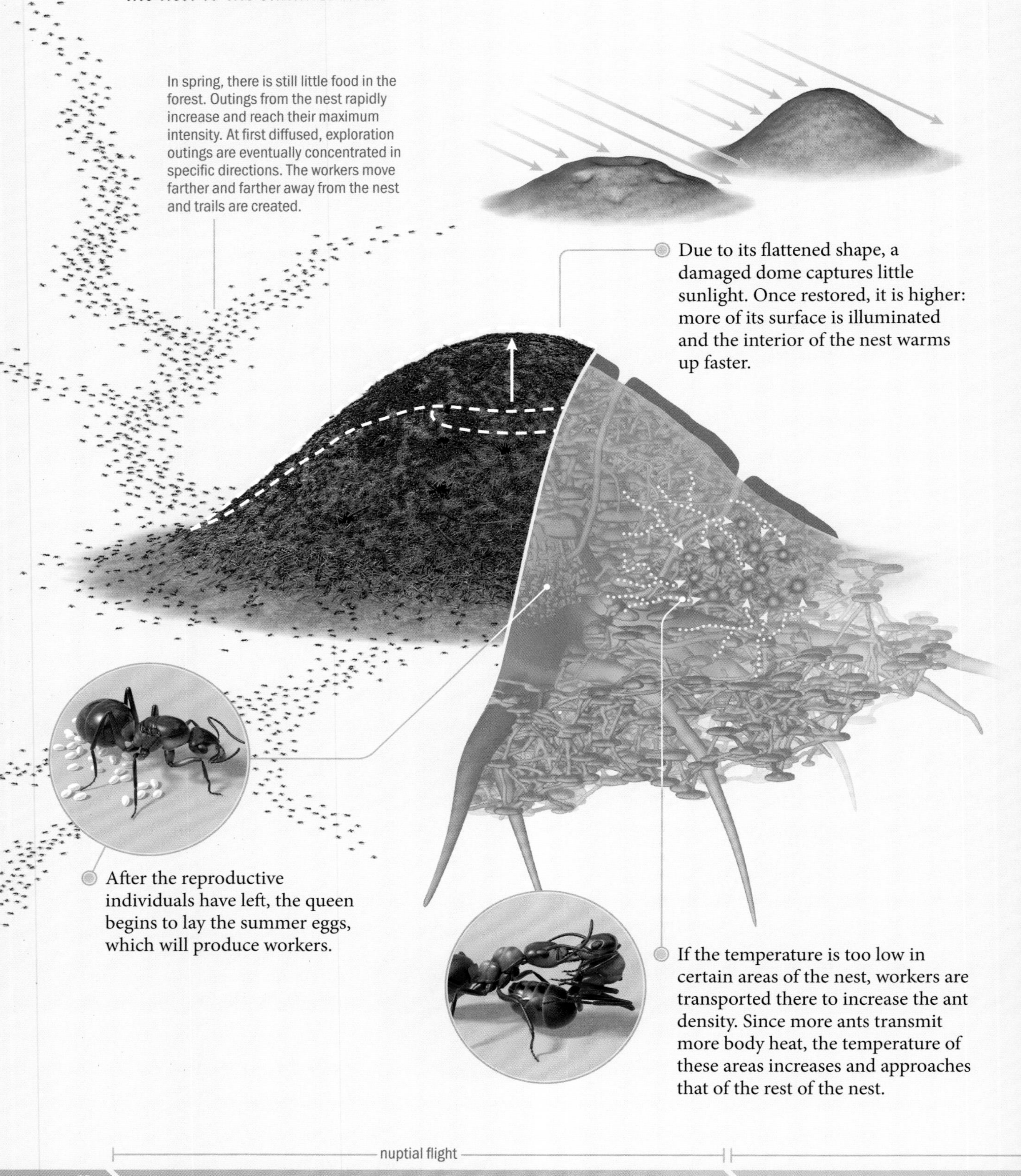

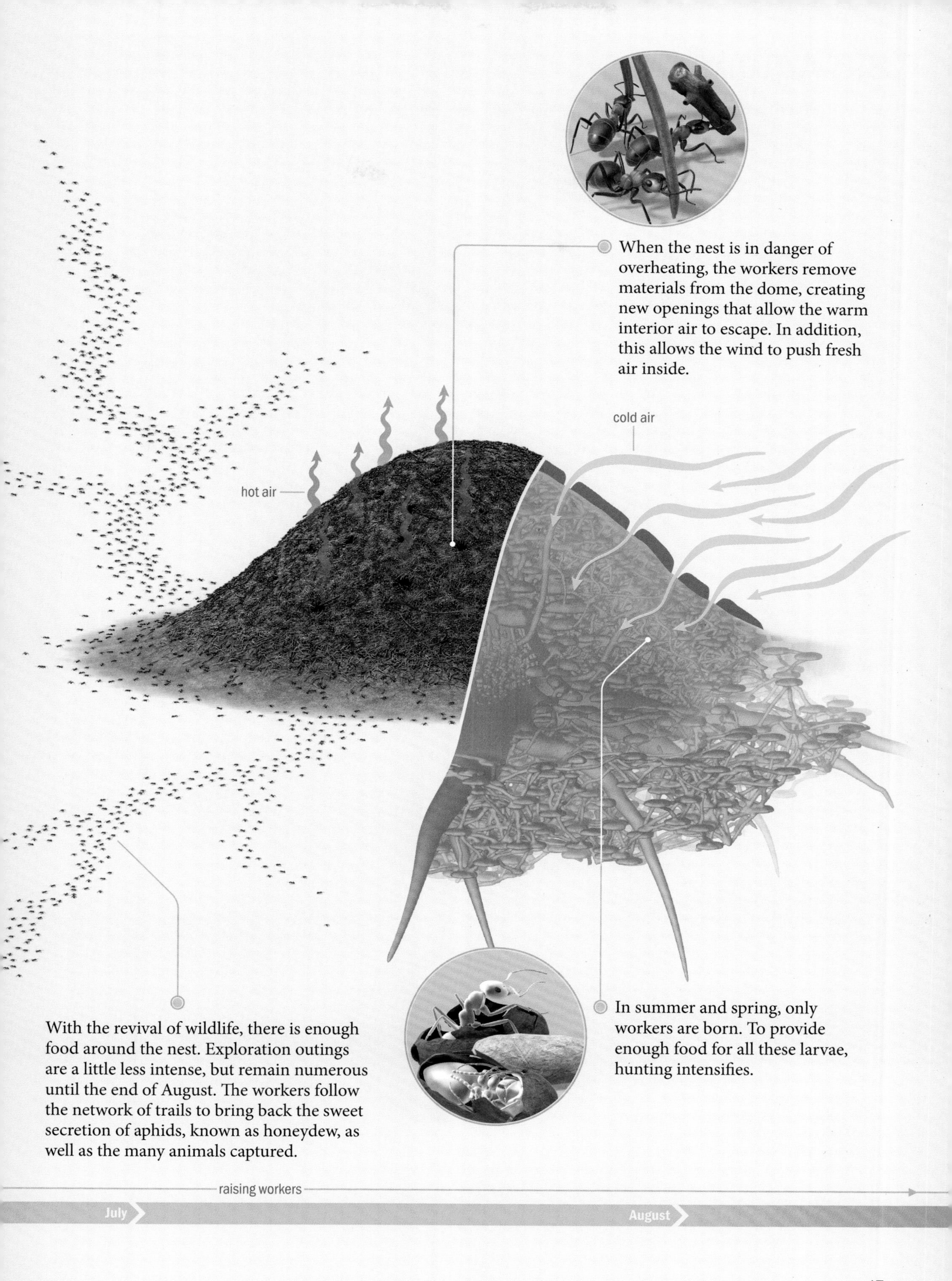

When the nest is in danger of overheating, the workers remove materials from the dome, creating new openings that allow the warm interior air to escape. In addition, this allows the wind to push fresh air inside.

With the revival of wildlife, there is enough food around the nest. Exploration outings are a little less intense, but remain numerous until the end of August. The workers follow the network of trails to bring back the sweet secretion of aphids, known as honeydew, as well as the many animals captured.

In summer and spring, only workers are born. To provide enough food for all these larvae, hunting intensifies.

3.5 Autumn

Toward the end of the season, the colony gets ready for winter rest and the nest is prepared for the cold.

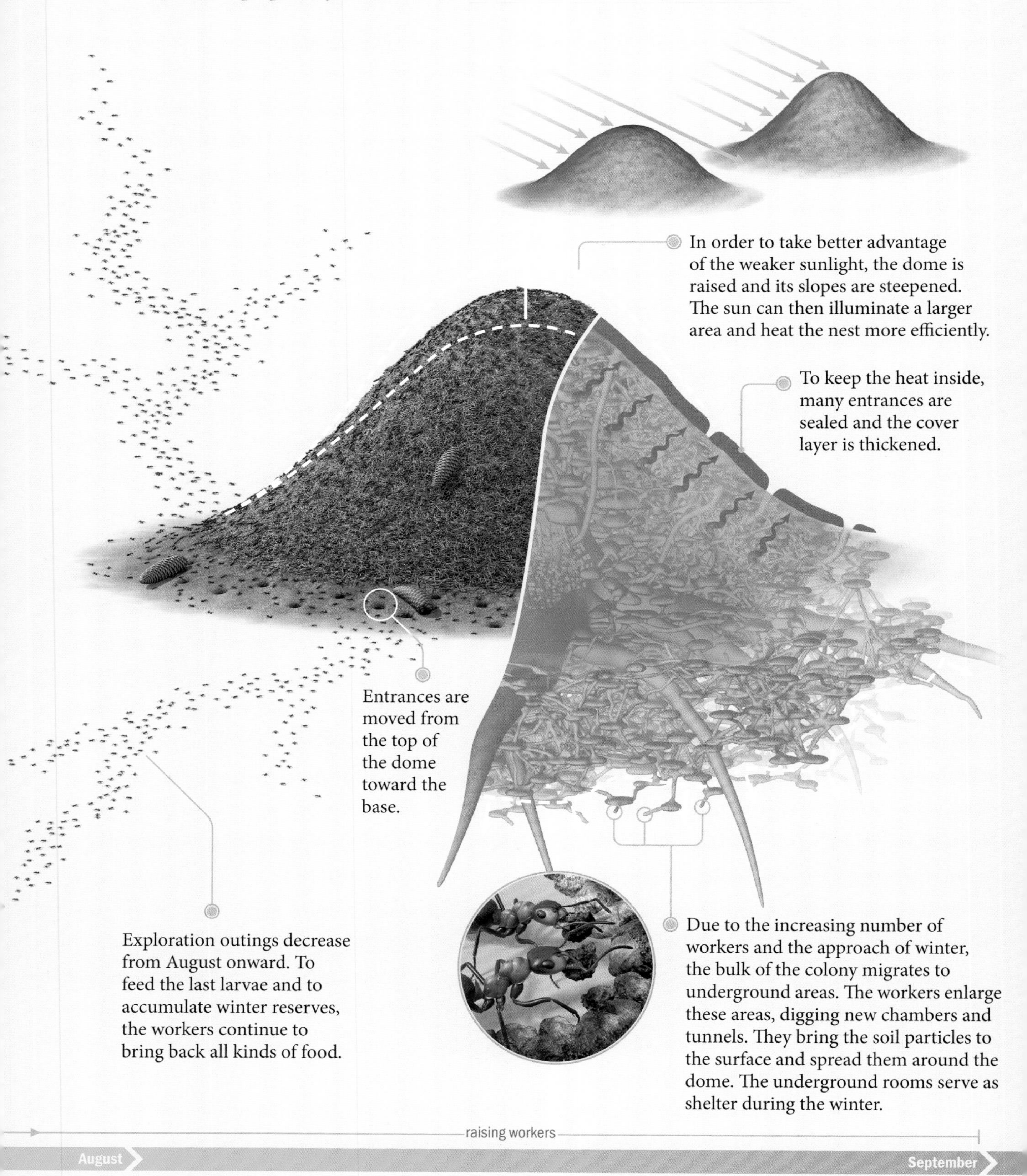

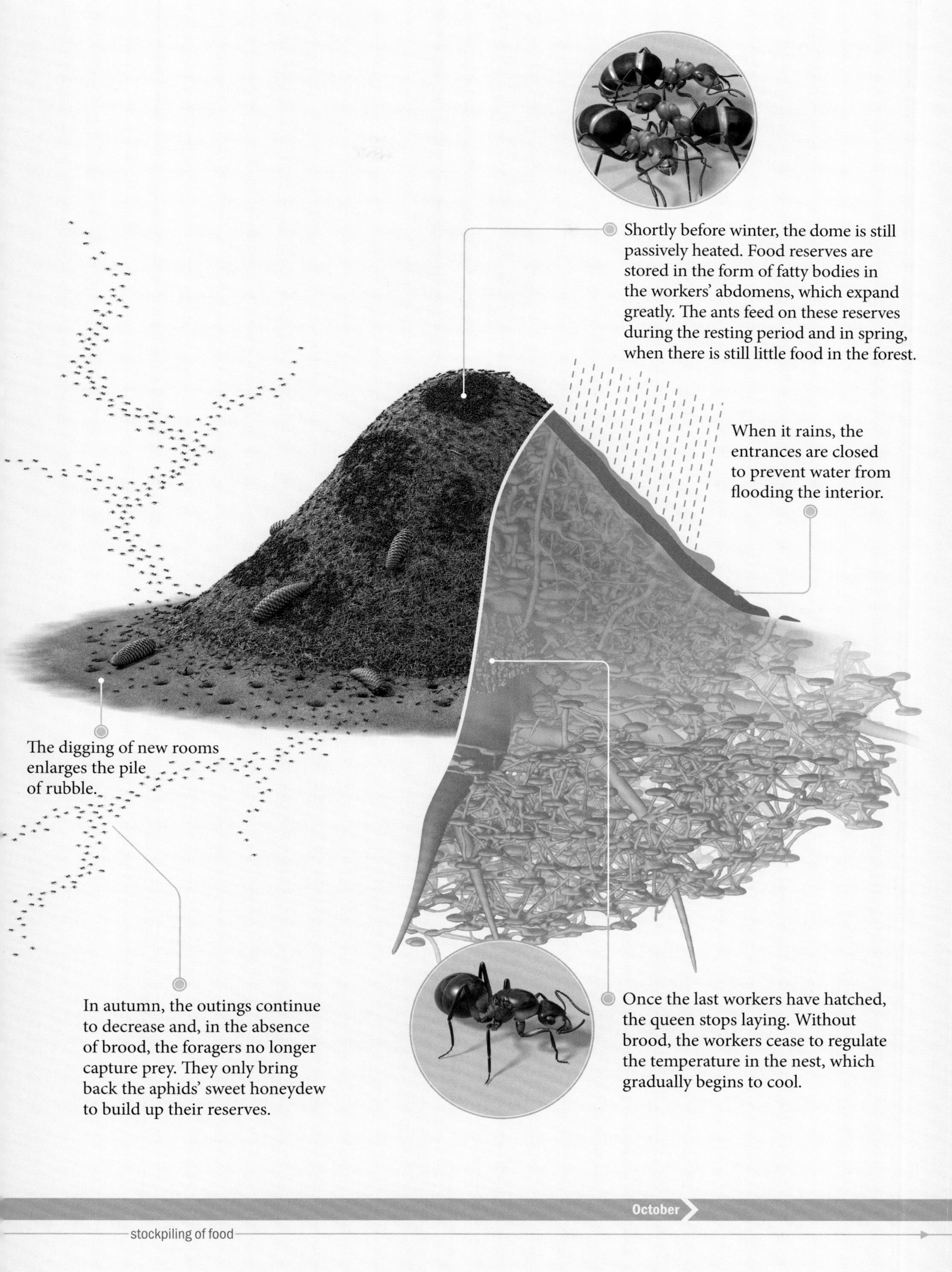
Shortly before winter, the dome is still passively heated. Food reserves are stored in the form of fatty bodies in the workers' abdomens, which expand greatly. The ants feed on these reserves during the resting period and in spring, when there is still little food in the forest.
When it rains, the entrances are closed to prevent water from flooding the interior.
The digging of new rooms enlarges the pile of rubble.
In autumn, the outings continue to decrease and, in the absence of brood, the foragers no longer capture prey. They only bring back the aphids' sweet honeydew to build up their reserves.
Once the last workers have hatched, the queen stops laying. Without brood, the workers cease to regulate the temperature in the nest, which gradually begins to cool.
October
stockpiling of food

3.6 Winter

In winter, the ants cease all their activities. To protect themselves from the cold and frost, they retreat underground, to the bottom of the nest.

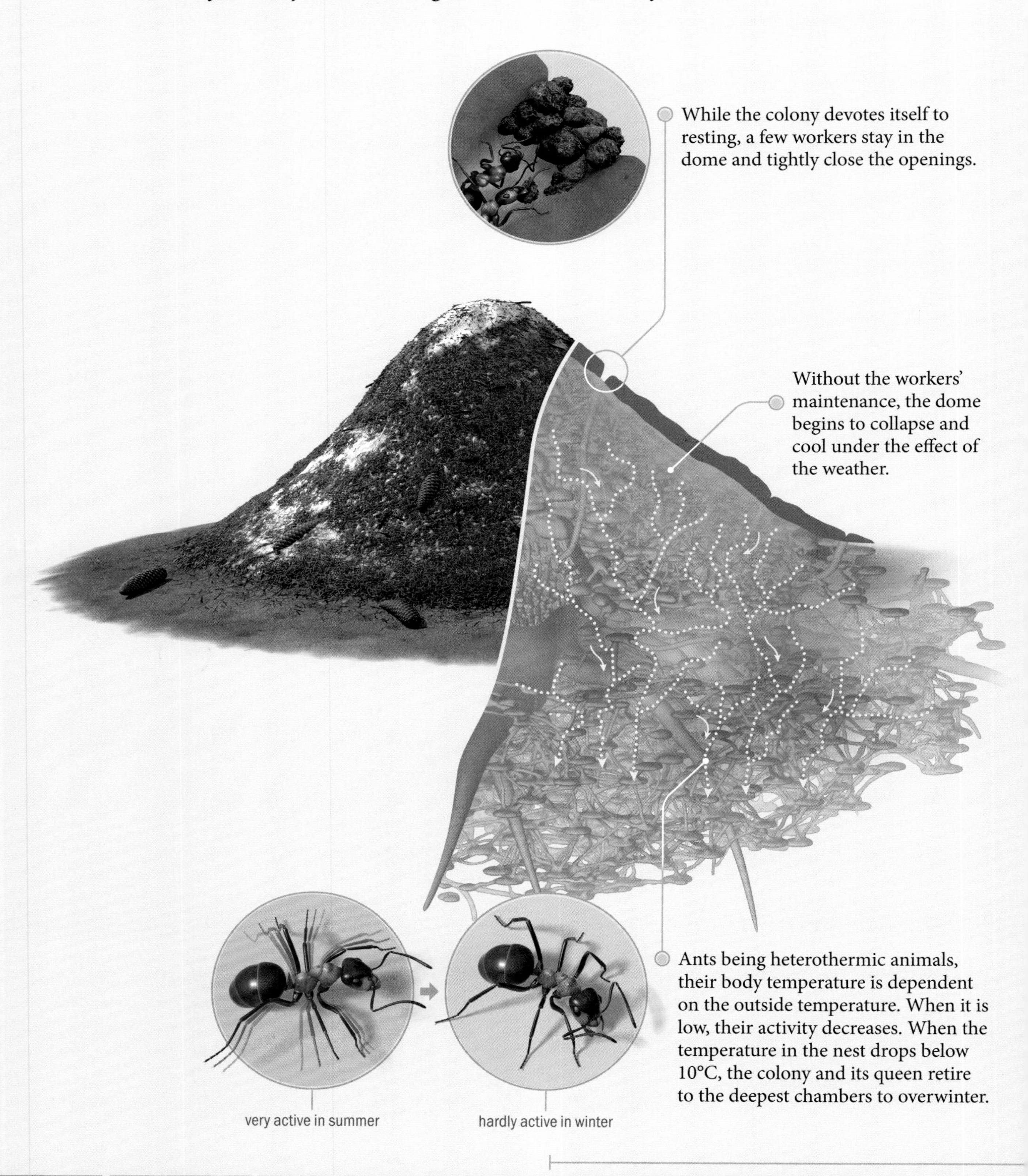

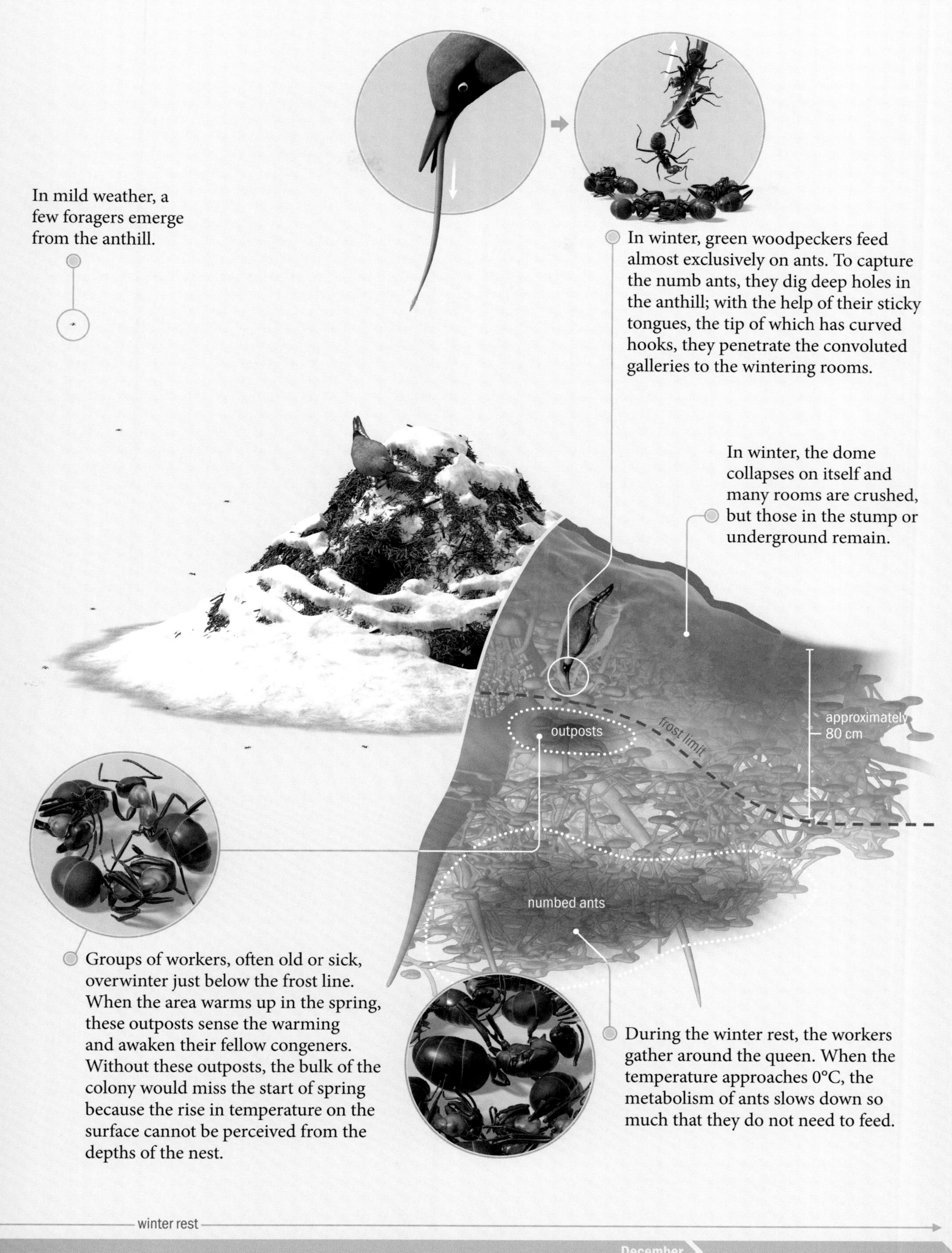
In mild weather, a few foragers emerge from the anthill.
In winter, green woodpeckers feed almost exclusively on ants. To capture the numb ants, they dig deep holes in the anthill; with the help of their sticky tongues, the tip of which has curved hooks, they penetrate the convoluted galleries to the wintering rooms.
In winter, the dome collapses on itself and many rooms are crushed, but those in the stump or underground remain.
outposts
frost limit
approximately 80 cm
numbed ants
Groups of workers, often old or sick, overwinter just below the frost line. When the area warms up in the spring, these outposts sense the warming and awaken their fellow congeners. Without these outposts, the bulk of the colony would miss the start of spring because the rise in temperature on the surface cannot be perceived from the depths of the nest.
During the winter rest, the workers gather around the queen. When the temperature approaches 0°C, the metabolism of ants slows down so much that they do not need to feed.
winter rest
December

4.1 Forest exploration

Foragers are constantly looking for food in the vicinity of the anthill.

2 Outside the nest, the scouts orient themselves by taking in the position of the sun and some noteworthy trees in the surrounding forest.
3 Along the way, they take advantage of natural trails, like fallen branches, to progress faster.

4.2 Ants' enemies

Outside the anthill, the workers are exposed to numerous predators specialized in capturing ants.

③ Because of their full poison reservoir, red wood ants are inedible to woodpeckers. To make them consumable, it rubs them on its plumage. When it does this, the ants project their defensive secretion, completely emptying their glands, and then the woodpecker can swallow them. In addition, the formic acid kills or drives away parasites from its feathers. Other birds, such as blackbirds and thrushes, take care of their plumage in the same way.
④ Red wood ants are the woodpecker's primary food, and it spends most of its time on the ground looking for them.
② The woodpecker quickly grabs several ants and keeps them in its beak.

The web spider of the *Cryptachaea riparia* species is an ant hunter. Hidden under a leaf, it waits for prey to get caught in one of its threads.
1
By wriggling, the ant makes the nest shake. The spider senses the vibrations, which alert it.
4
2
An ant bumps into a thread and gets trapped by the sticky drops that dot it.

6 The spider drags the ant to its nest. Wrapped in a silk cocoon, it serves as a food reserve for the spider and its larvae.
5 The spider wraps the ant in its thread to immobilize it; then it bites the nape of its neck to paralyze it with its venom.
7 The spider disposes of the ant's dried remains by dropping them on the ground.
3 Struggling to free itself, the ant detaches the thread from the ground. It is lifted into the air and becomes entangled in the thread.

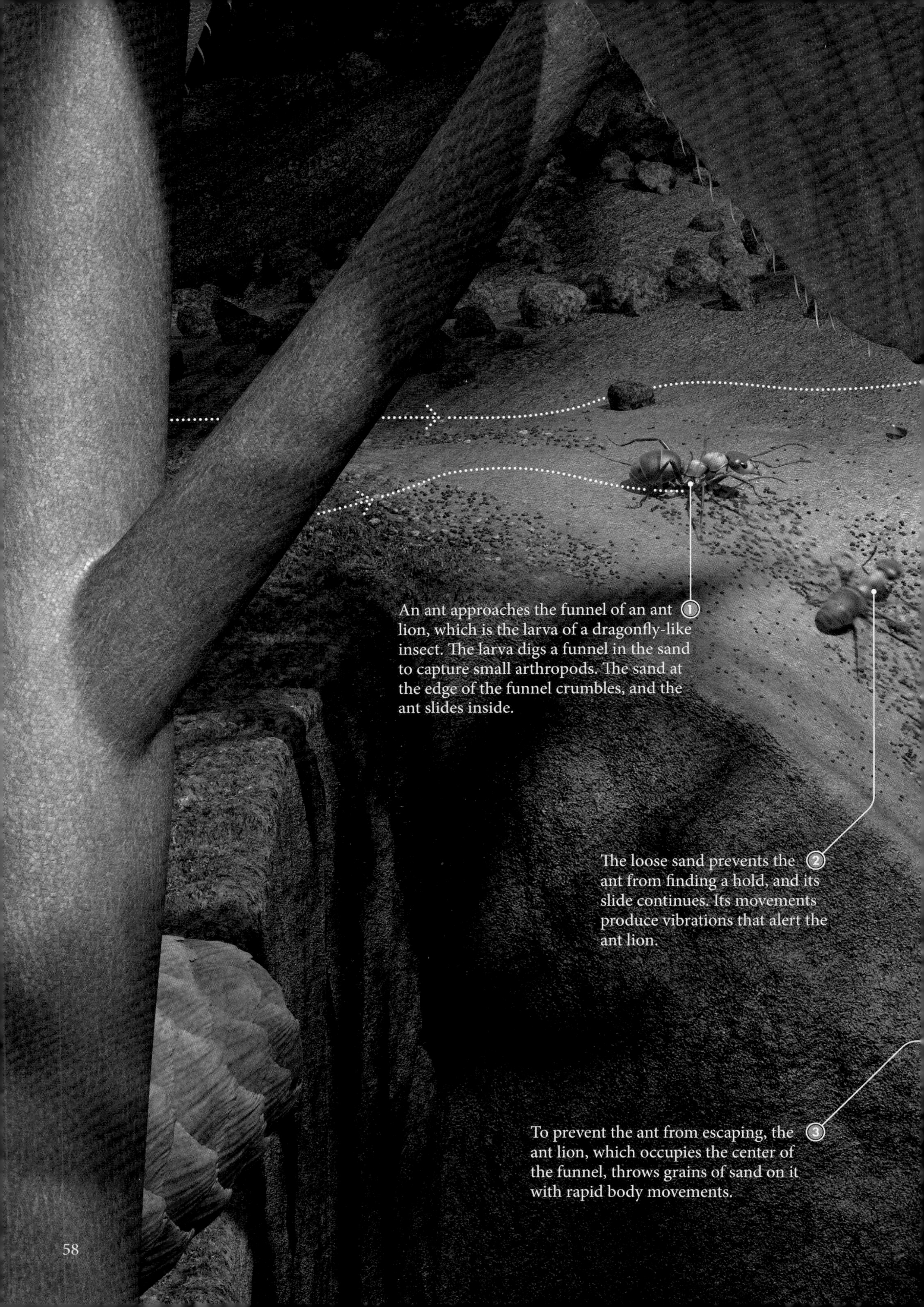

1 An ant approaches the funnel of an ant lion, which is the larva of a dragonfly-like insect. The larva digs a funnel in the sand to capture small arthropods. The sand at the edge of the funnel crumbles, and the ant slides inside.
2 The loose sand prevents the ant from finding a hold, and its slide continues. Its movements produce vibrations that alert the ant lion.
3 To prevent the ant from escaping, the ant lion, which occupies the center of the funnel, throws grains of sand on it with rapid body movements.

6 The ant lion throws the ant's dried remains out of the funnel.
4 As soon as the ant is within reach, the ant lion grabs it with its mandibles and paralyzes it with its venom.
5 The ant lion pulls its prey under the sand. It sucks the liquid from its body and completely empties it. When an ant lion funnel is near an anthill, ants are its primary food.

4.3 Foraging

The scout has found food. She must now find her way back to the nest and lead other foragers to the food source.

① The scout tastes the food. If it suits her, she fills her crop with it. The more she likes the food, the more she ingests.

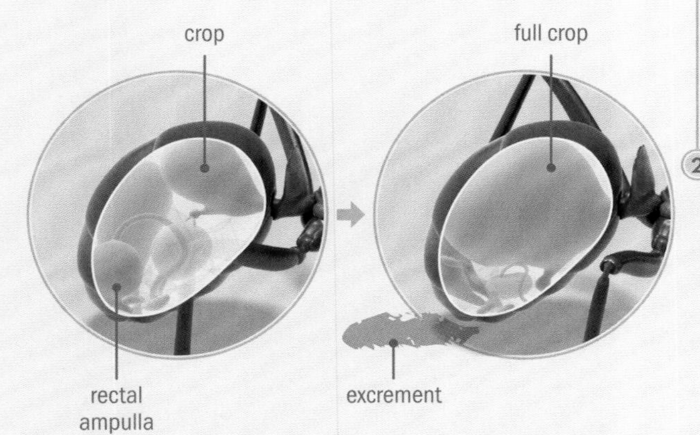

② The ant's crop expands until it fills much of the abdomen, compressing other organs. The rectal ampulla is also compressed and releases excrement. The amount defecated depends on the amount of food ingested.

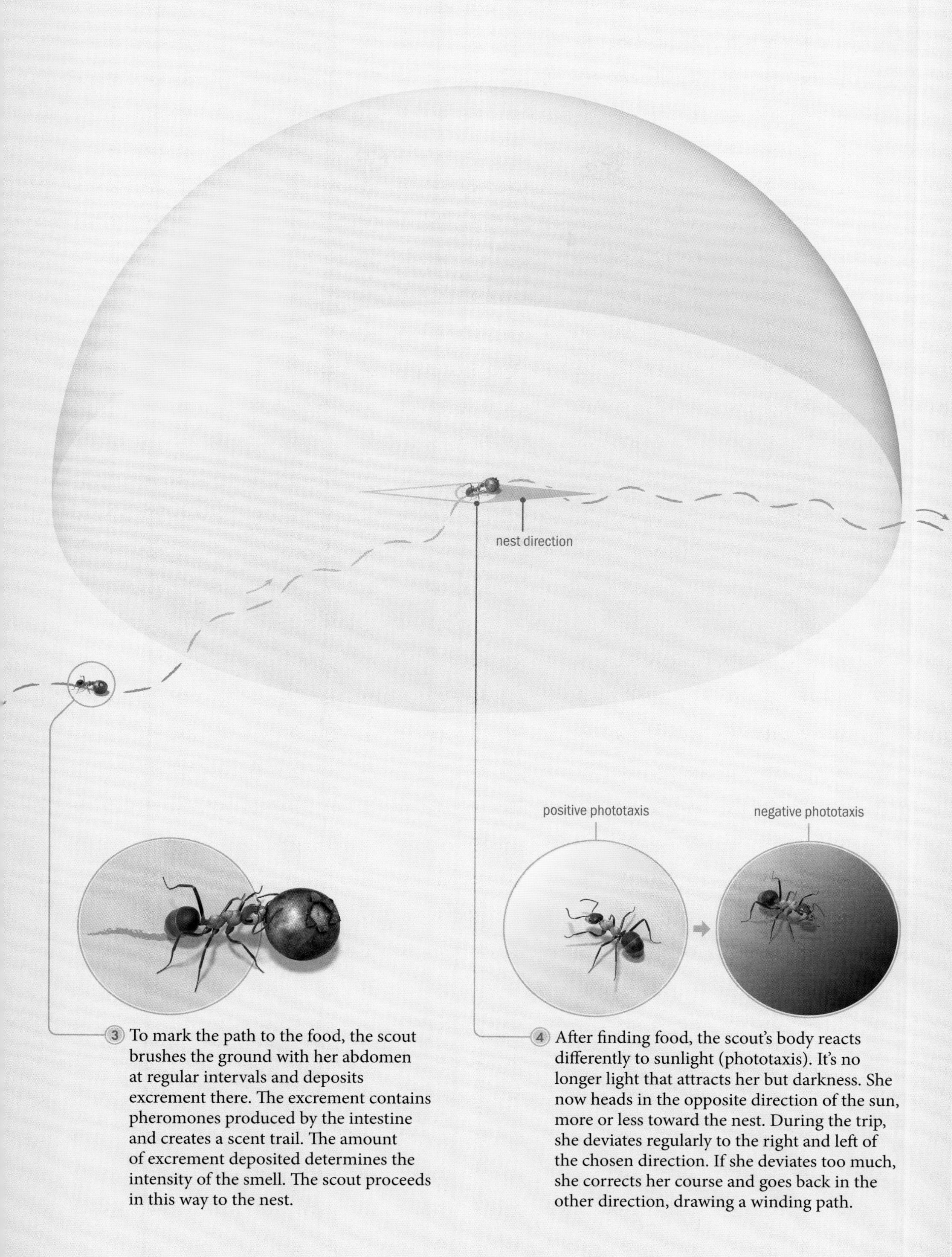

3 To mark the path to the food, the scout brushes the ground with her abdomen at regular intervals and deposits excrement there. The excrement contains pheromones produced by the intestine and creates a scent trail. The amount of excrement deposited determines the intensity of the smell. The scout proceeds in this way to the nest.

4 After finding food, the scout's body reacts differently to sunlight (phototaxis). It's no longer light that attracts her but darkness. She now heads in the opposite direction of the sun, more or less toward the nest. During the trip, she deviates regularly to the right and left of the chosen direction. If she deviates too much, she corrects her course and goes back in the other direction, drawing a winding path.

4.4 Visual orientation

On the return journey, the worker's two complex eyes give her a rough picture of her surroundings.

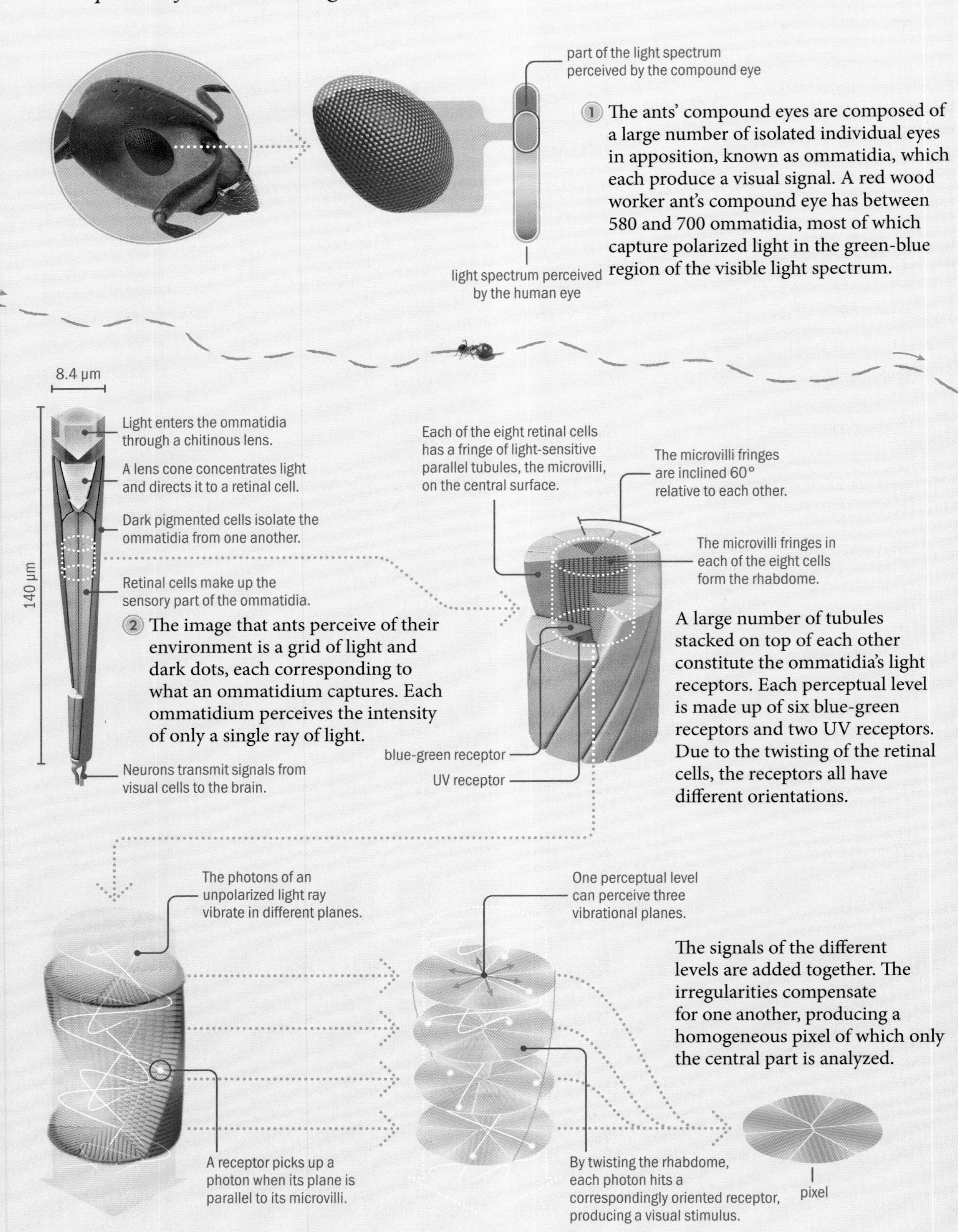

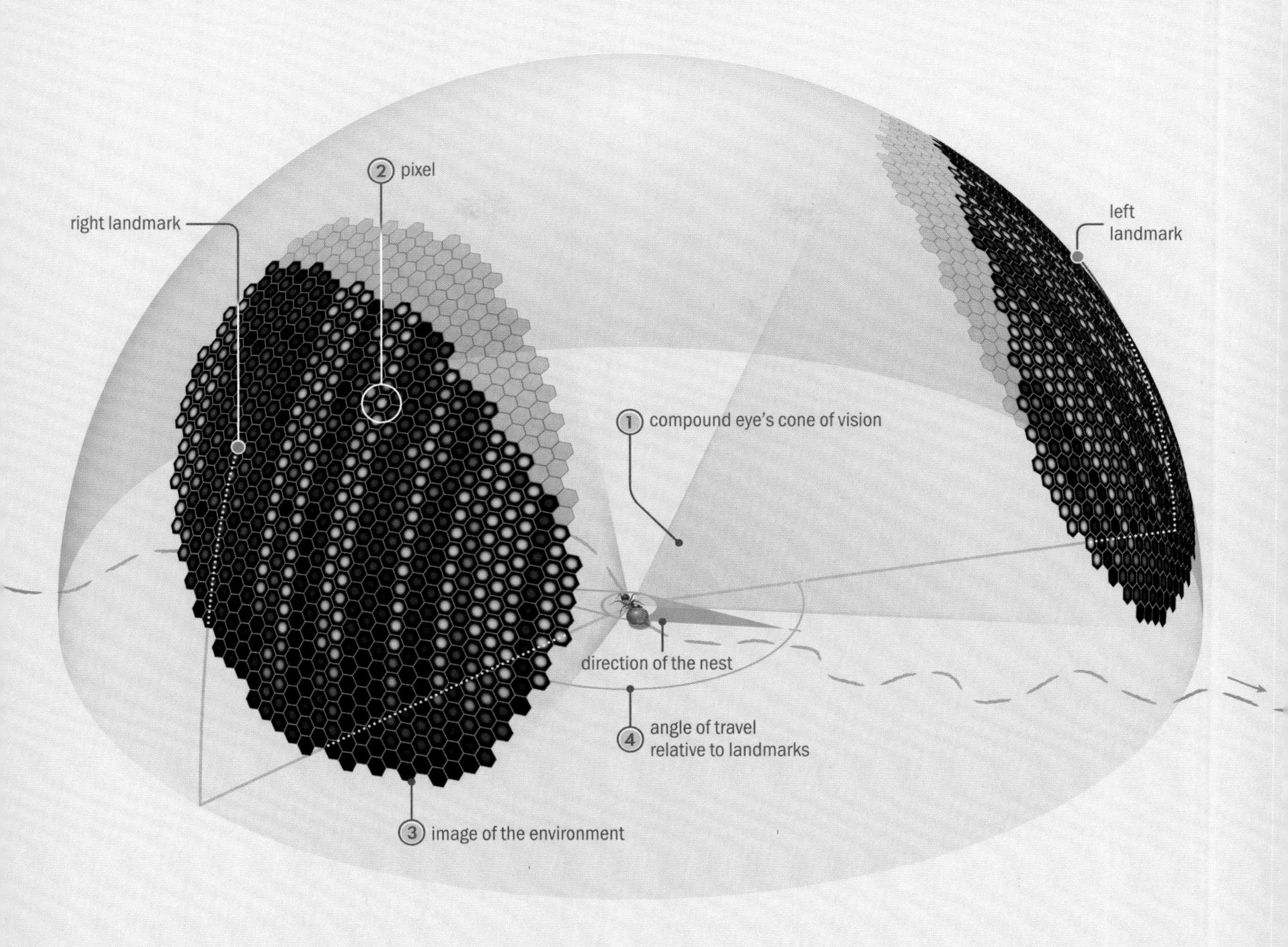

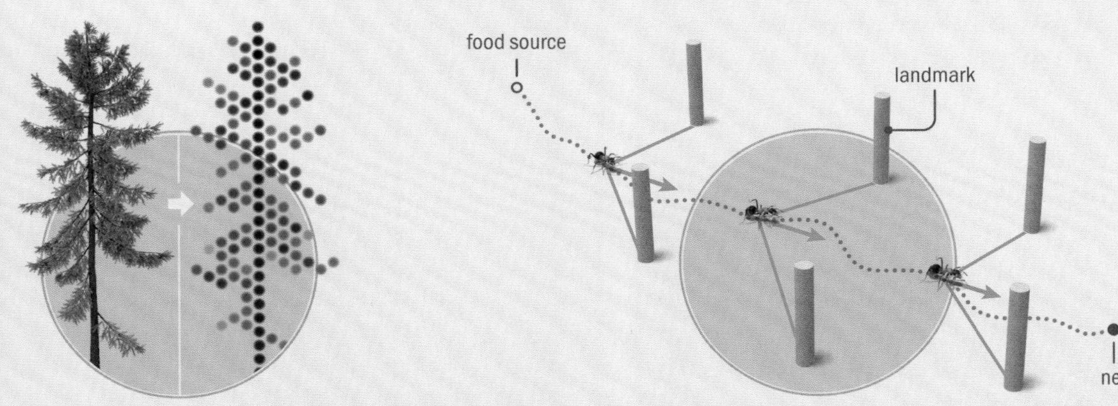

3 In the brain, the ommatidia's pixels are assembled to form a rough vertical image. The image of the forest is thus reduced to vertical silhouettes of the trunks and foliage patterns. The trunks' silhouettes serve as the scout's orientation landmarks.

4 The scout has recognized the nest's surroundings from previous outings. She memorized the trees and light sources, as well as the angle of her route. On the way back, she recognizes these landmarks and remembers the corresponding angle, and she takes the correct direction to return to the anthill.

4.5 Celestial orientation

The position of the sun and its light polarization are landmarks that complete the visual orientation.

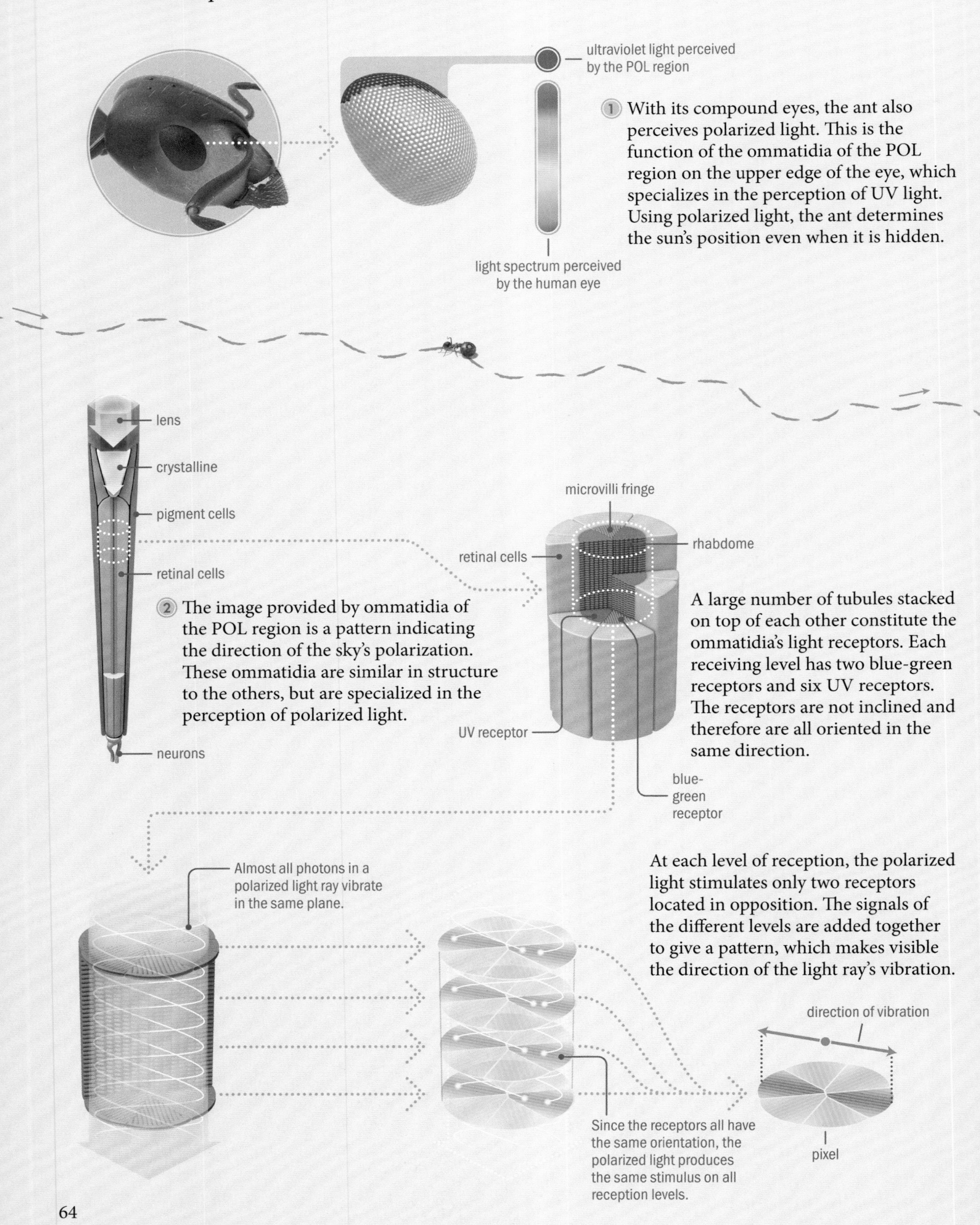

① With its compound eyes, the ant also perceives polarized light. This is the function of the ommatidia of the POL region on the upper edge of the eye, which specializes in the perception of UV light. Using polarized light, the ant determines the sun's position even when it is hidden.

② The image provided by ommatidia of the POL region is a pattern indicating the direction of the sky's polarization. These ommatidia are similar in structure to the others, but are specialized in the perception of polarized light.

A large number of tubules stacked on top of each other constitute the ommatidia's light receptors. Each receiving level has two blue-green receptors and six UV receptors. The receptors are not inclined and therefore are all oriented in the same direction.

At each level of reception, the polarized light stimulates only two receptors located in opposition. The signals of the different levels are added together to give a pattern, which makes visible the direction of the light ray's vibration.

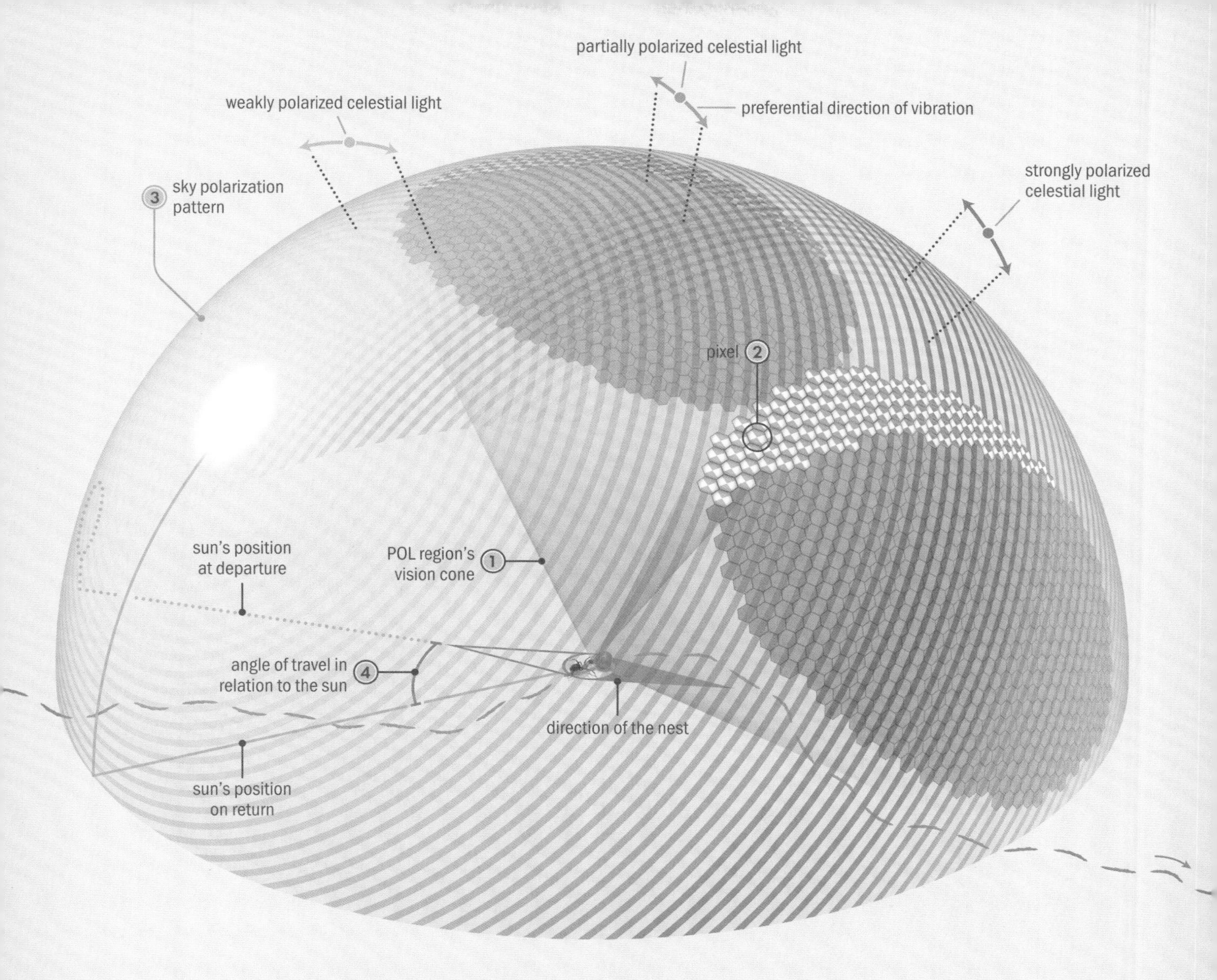

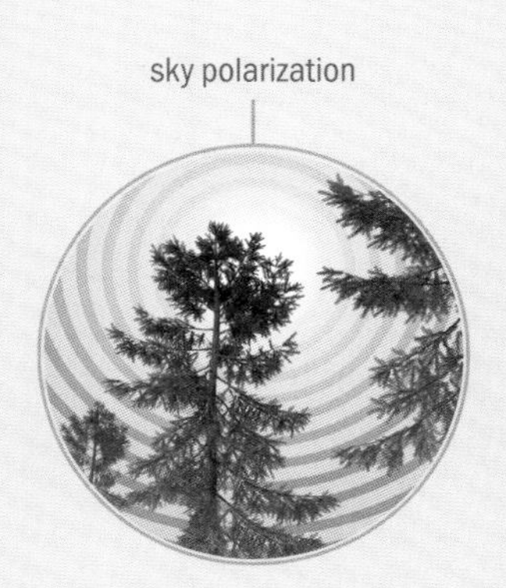

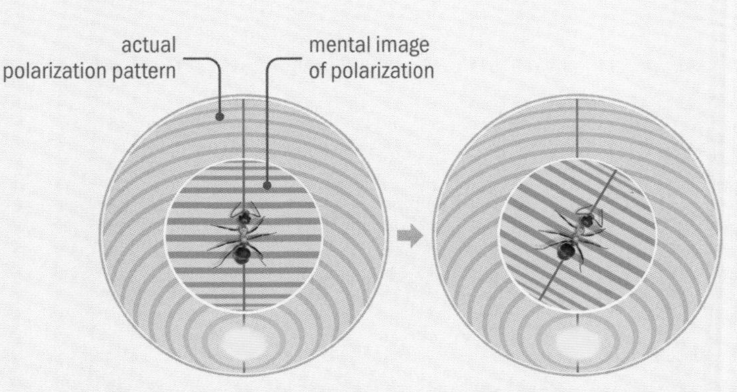

3 The sky polarization pattern is produced by the refraction and reflection of sunlight in the earth's atmosphere. Regardless of their angle of incidence, light rays are more or less polarized. The preferential direction of polarized light vibration is always perpendicular to the direction of the sun. The directions of the different light rays' polarization thus create a pattern of concentric circles around the sun, allowing the ant to orient itself.

4 The scout compares the sky polarization pattern with a simplified image stored in her brain. The degree of correspondence between the two images allows her to know the angle of her course in relation to the sun. This faculty serves as a compass to determine the direction of the nest: when leaving the nest, she goes toward the sun; on her return, she turns her back on it. But since the sun has changed its position in the meantime, she follows a certain angle with respect to it in order to take the right path.

4.6 Return to the nest

Arriving near the nest, the scout meets members of her colony. Their scent helps her find her way back.

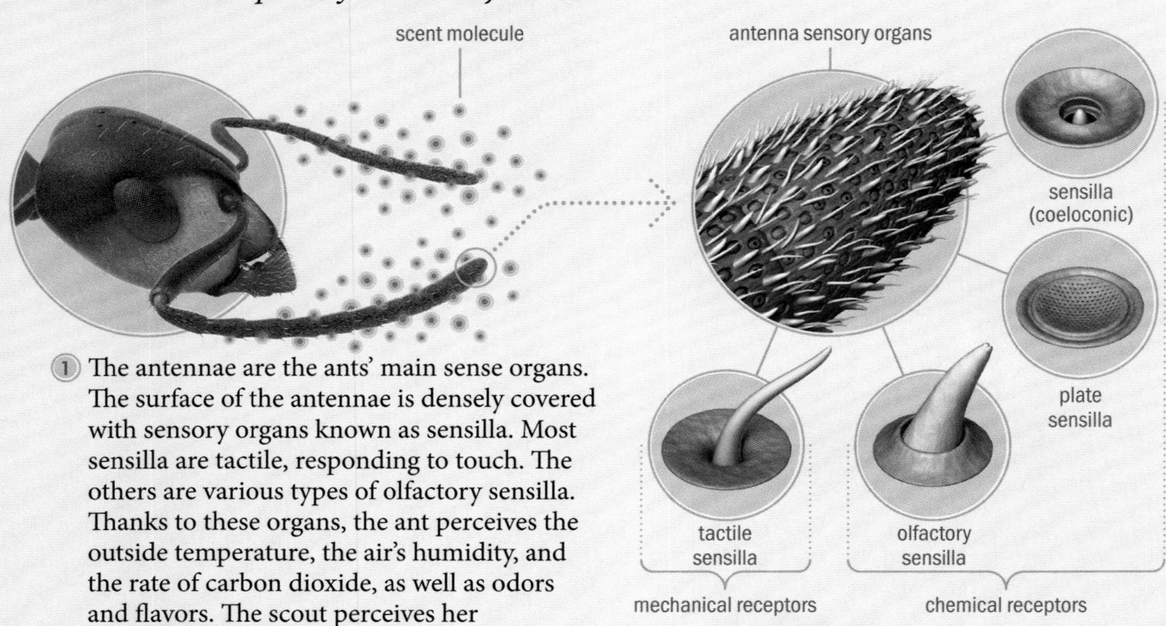

① The antennae are the ants' main sense organs. The surface of the antennae is densely covered with sensory organs known as sensilla. Most sensilla are tactile, responding to touch. The others are various types of olfactory sensilla. Thanks to these organs, the ant perceives the outside temperature, the air's humidity, and the rate of carbon dioxide, as well as odors and flavors. The scout perceives her congeners' scent and follows it.

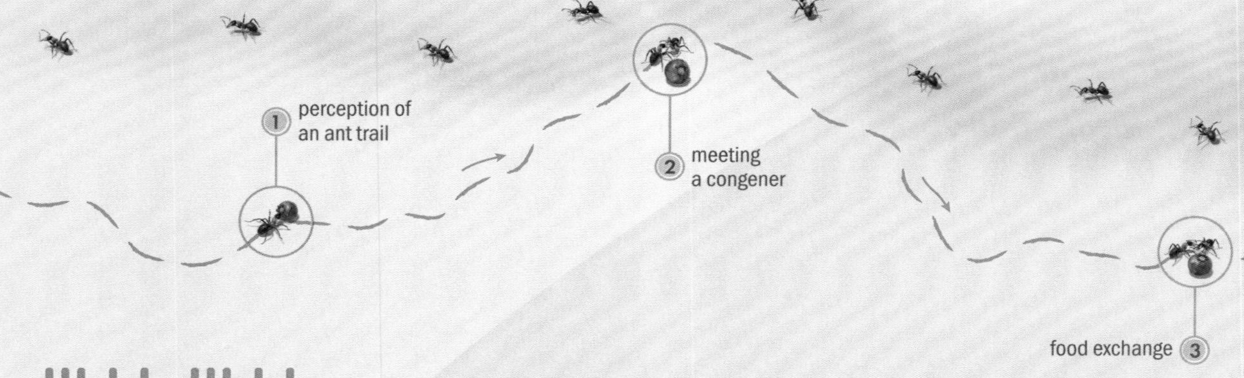

② The scent leads to a nearby trail, where she encounters another worker. The two ants feel each other and check each other's collective scent. If the two ants belong to the same nest, they separate and the scout follows the trail toward the nest.

③ When two workers from the same colony meet, they often exchange food, a behavior called trophallaxis. Encounters, checking each other's scent, and trophallaxis are repeated several times until the scout reaches the nest.

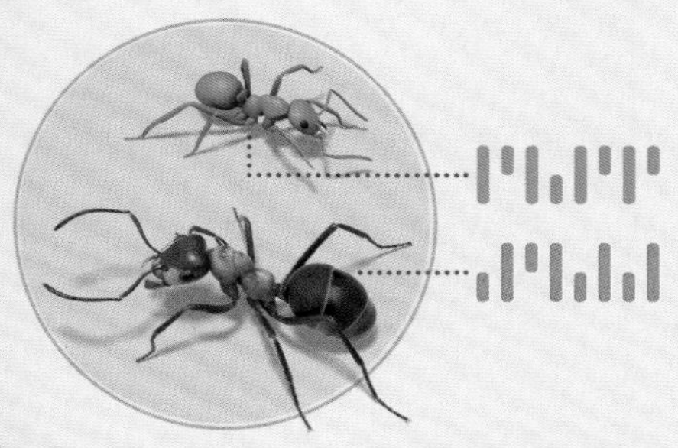

1 When the workers encounter a greater number of ants from a colony or a foreign species in unfamiliar territory, they adopt a defensive attitude and avoid them.

2 On familiar ground and in the presence of other members of their colony, the workers are more aggressive toward foreign ants, assuming a threatening posture and striving to drive them away.

3 In close proximity to the nest, the ants are very aggressive. Intruders are collectively attacked and killed before being transported to the nest, where they are used as food.

Once the scout arrives, she recruits other workers to lead them to the food source.

In order to exploit the food source, the scout needs help. Back at the nest, she informs her congeners of her discovery and recruits foragers.

The blueberry she brought back arouses the attention of the colony. The neighboring workers rush in and start tearing up the blueberry.

The scout strives to transmit to her fellow congeners the excitement of her discovery. She regurgitates a drop of blueberry pulp and offers it to another worker, whose head she taps.

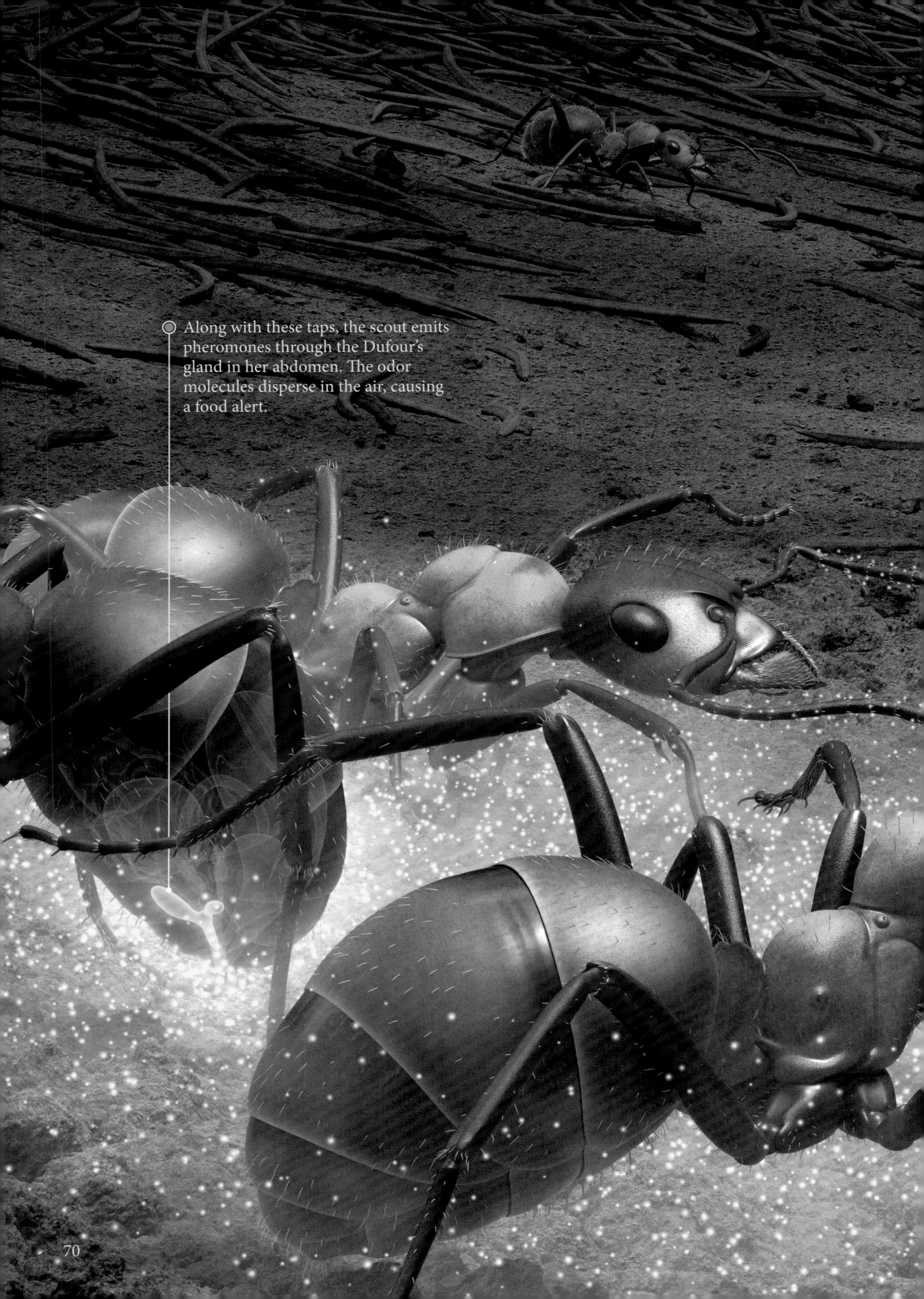
Along with these taps, the scout emits pheromones through the Dufour's gland in her abdomen. The odor molecules disperse in the air, causing a food alert.

Neighboring workers' olfactory receptors on their antennae perceive odorous molecules. If the concentration is sufficient, the scent incites the workers to leave the nest and to go in search of food.

The scent trail that the scout left on returning to the nest helps the foragers orient themselves and leads them to the food source.

The ants' two antennae provide them with a spatial perception of the scent, which allows them to follow it precisely.

The recruited foragers are generally older and experienced. When food is lacking in the anthill, young, inexperienced ants are also recruited. On their way to the food source, the foragers memorize the position of the trees in order to be able to orient themselves later without the help of the scent trail.

At the end of the scent trail, the foragers discover the blueberries. They fill their crops and set off for the return journey. The smallest berries are carried in groups to the nest.

5.2 Birth of an ant trail

Ant trails usually lead to abundant food sources.

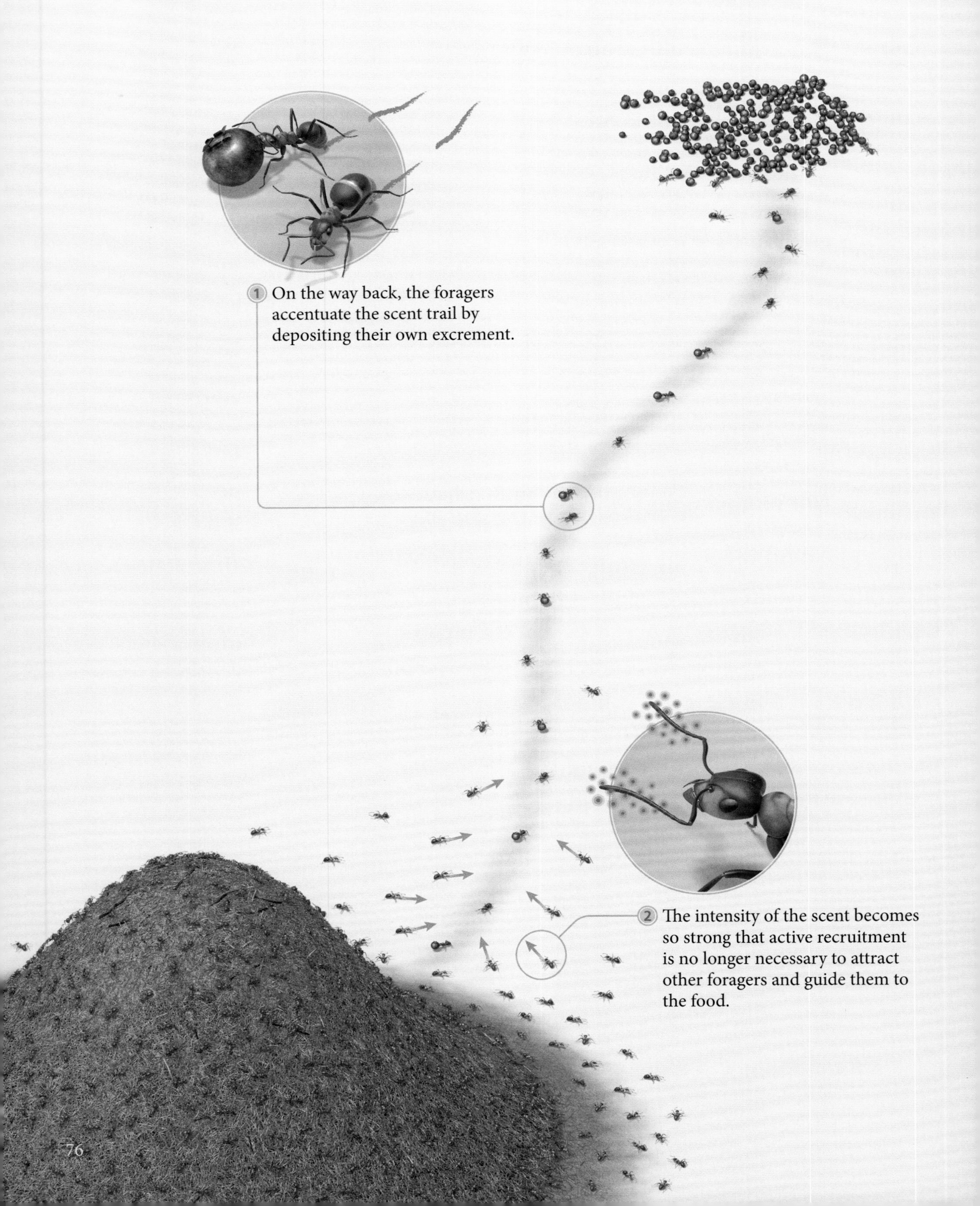

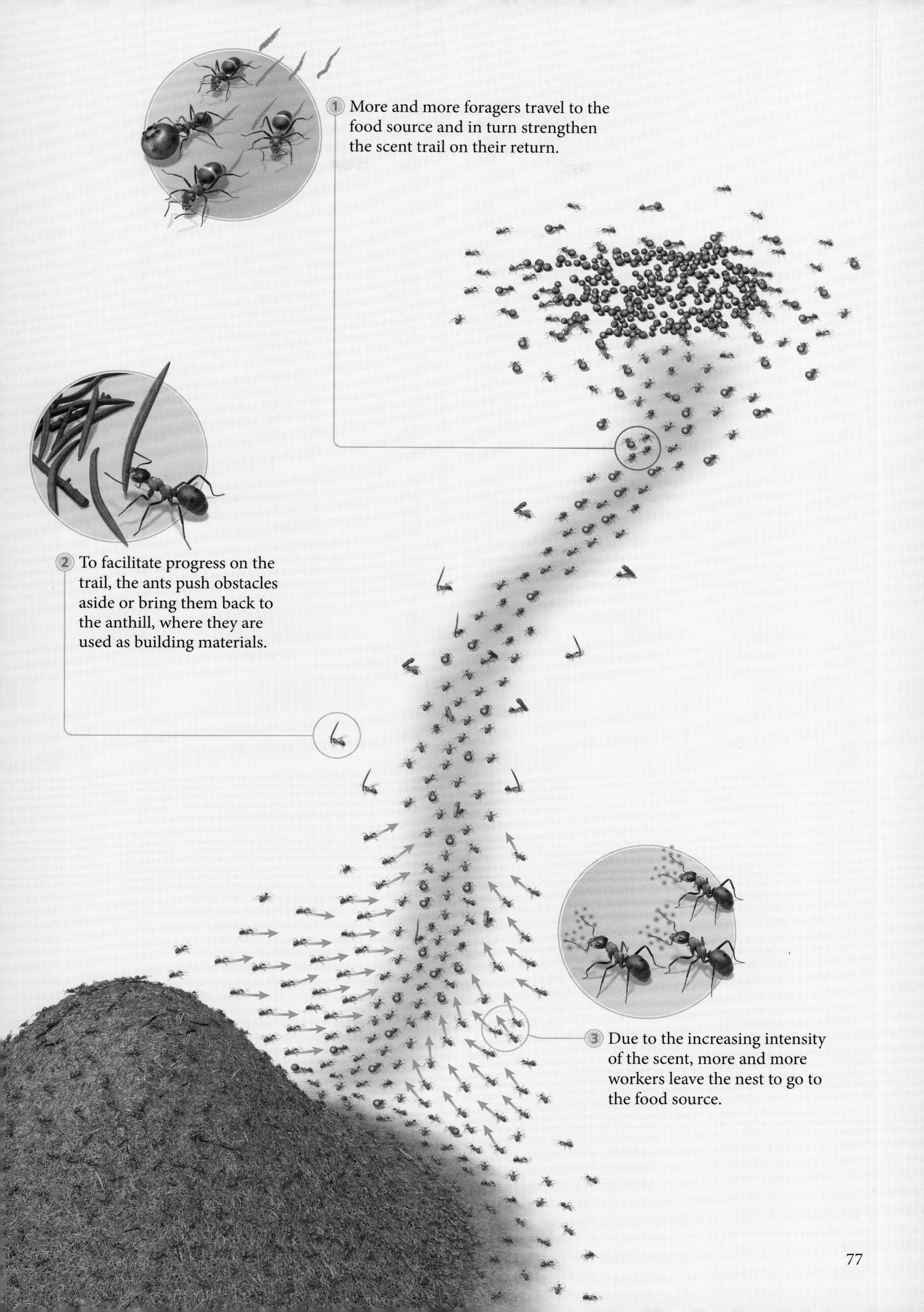
1 More and more foragers travel to the food source and in turn strengthen the scent trail on their return.
2 To facilitate progress on the trail, the ants push obstacles aside or bring them back to the anthill, where they are used as building materials.
3 Due to the increasing intensity of the scent, more and more workers leave the nest to go to the food source.

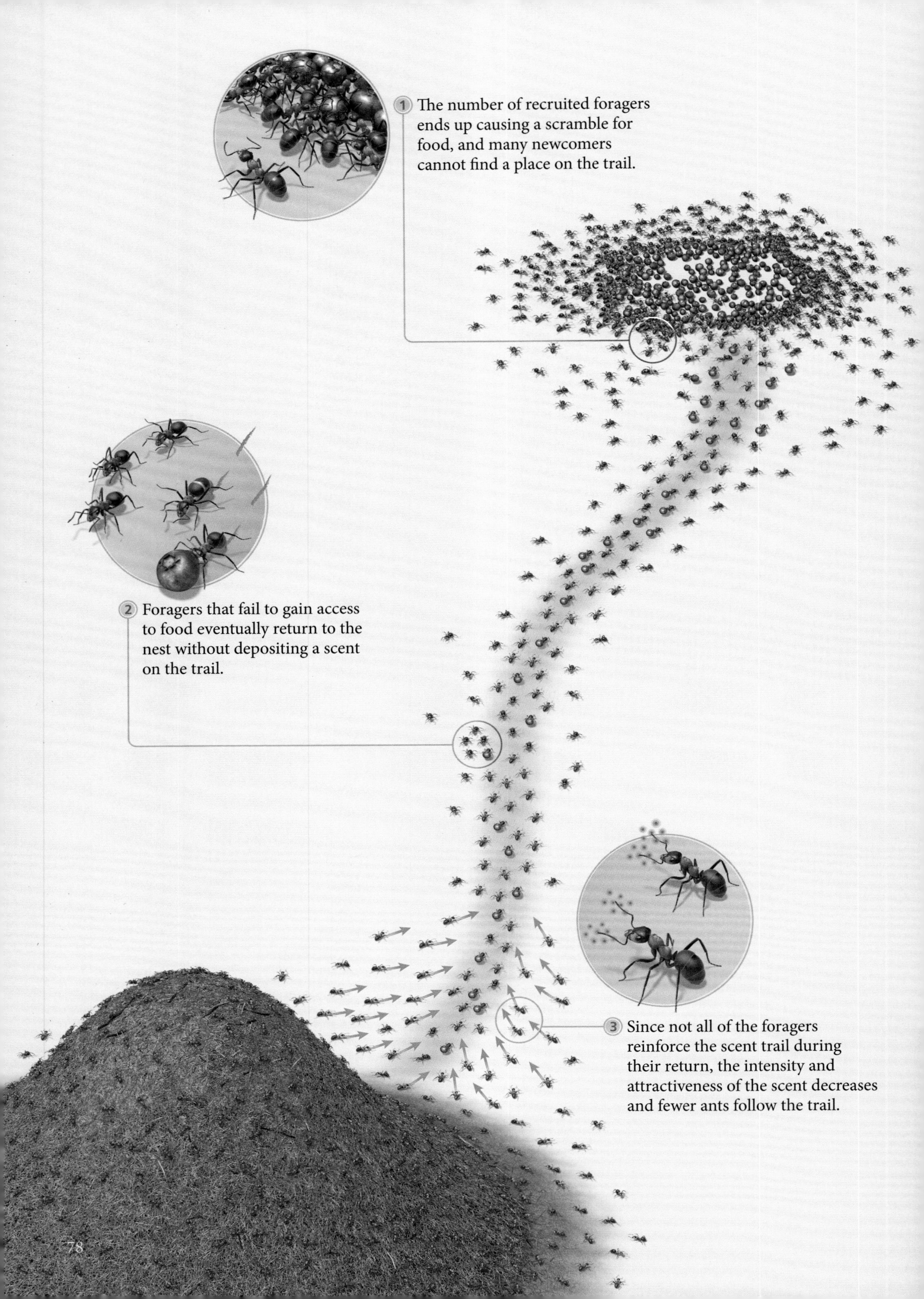
1 The number of recruited foragers ends up causing a scramble for food, and many newcomers cannot find a place on the trail.
2 Foragers that fail to gain access to food eventually return to the nest without depositing a scent on the trail.
3 Since not all of the foragers reinforce the scent trail during their return, the intensity and attractiveness of the scent decreases and fewer ants follow the trail.

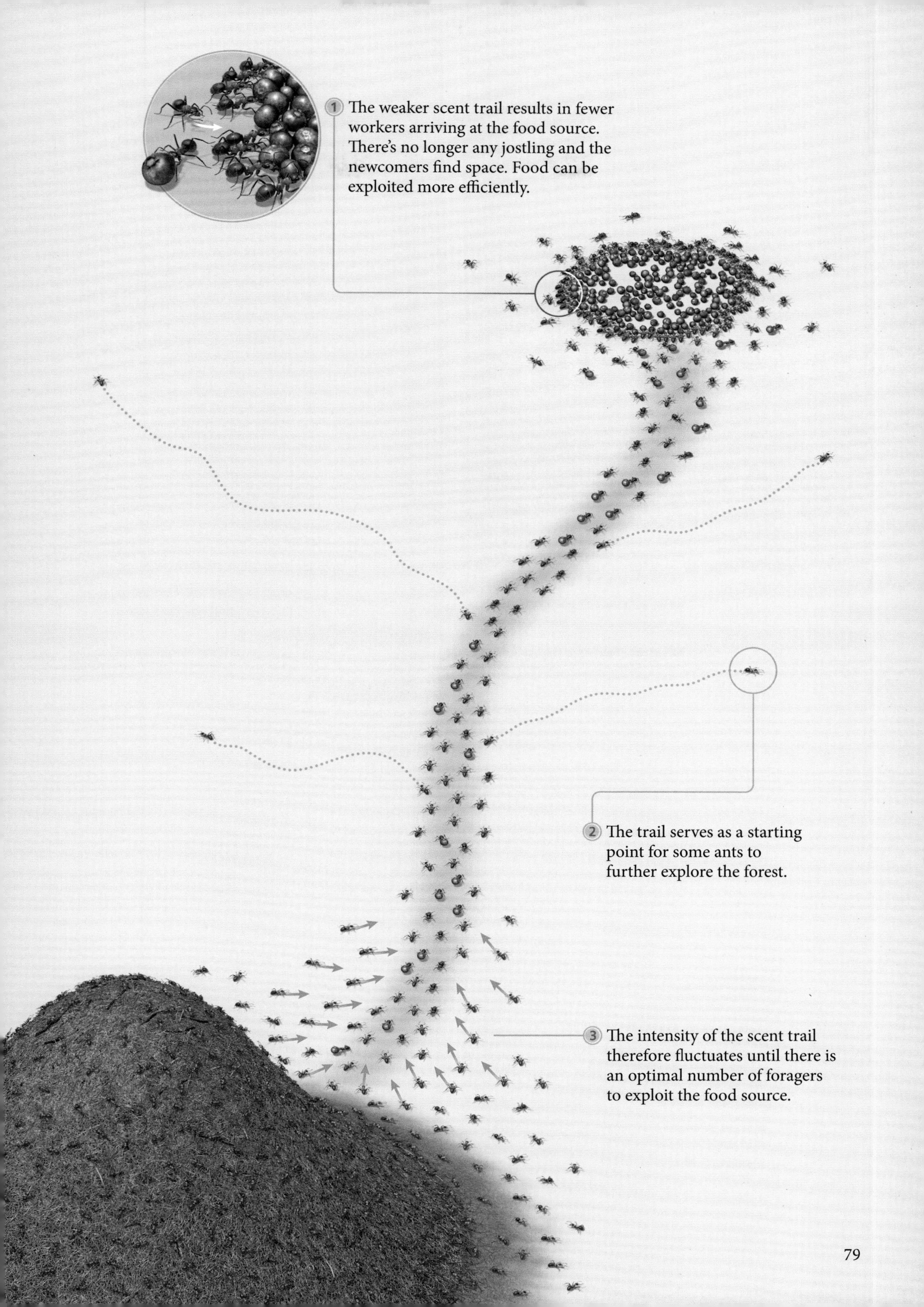
1 The weaker scent trail results in fewer workers arriving at the food source. There's no longer any jostling and the newcomers find space. Food can be exploited more efficiently.
2 The trail serves as a starting point for some ants to further explore the forest.
3 The intensity of the scent trail therefore fluctuates until there is an optimal number of foragers to exploit the food source.

1 The food source begins to deplete, and the foragers adapt to the lower quantities by depositing a less intense scent trail. The attractiveness of the trail decreases even more and fewer ants go to the food.

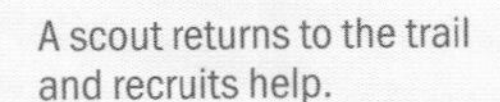

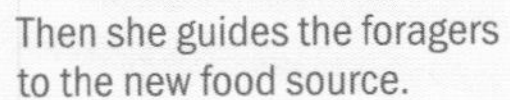

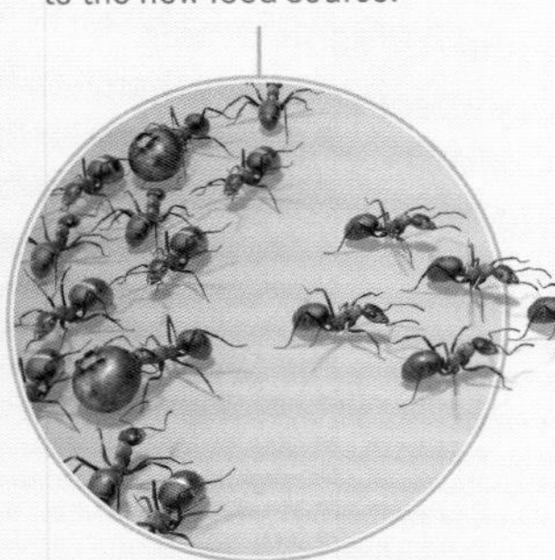

2 When a scout has found another food source, she recruits foragers and leads them there. If the new source is promising, the process repeats itself and another trail is born, an offshoot of the first.

At the branching level, it is the scent trail's intensity that determines which direction is chosen most often.

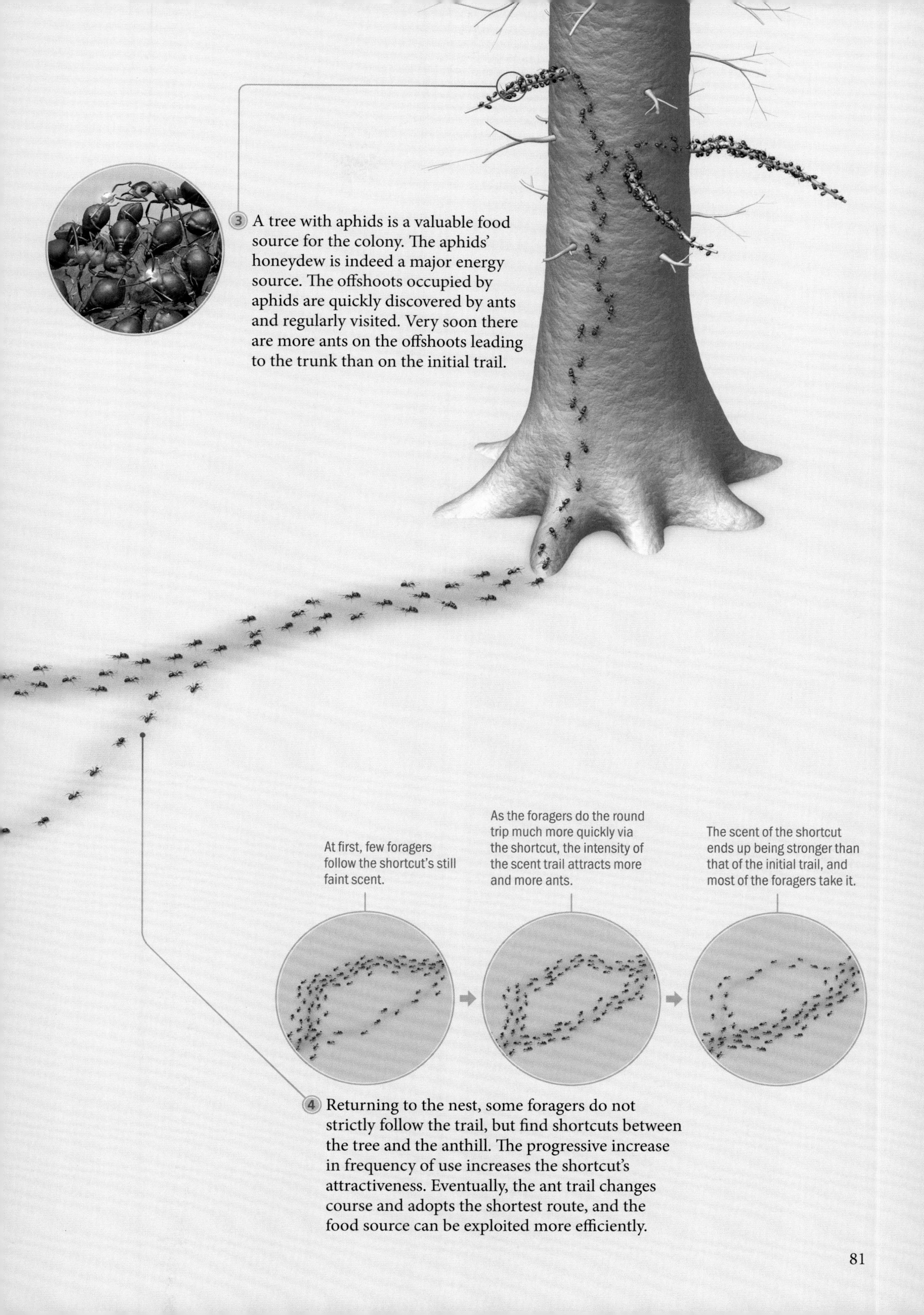
3 A tree with aphids is a valuable food source for the colony. The aphids' honeydew is indeed a major energy source. The offshoots occupied by aphids are quickly discovered by ants and regularly visited. Very soon there are more ants on the offshoots leading to the trunk than on the initial trail.
At first, few foragers follow the shortcut's still faint scent.
As the foragers do the round trip much more quickly via the shortcut, the intensity of the scent trail attracts more and more ants.
The scent of the shortcut ends up being stronger than that of the initial trail, and most of the foragers take it.
4 Returning to the nest, some foragers do not strictly follow the trail, but find shortcuts between the tree and the anthill. The progressive increase in frequency of use increases the shortcut's attractiveness. Eventually, the ant trail changes course and adopts the shortest route, and the food source can be exploited more efficiently.

A tree hosts several colonies of aphids that feed on its sap. Their sweet exudate, honeydew, is an important food source for red wood ants. To harvest the honeydew, the workers make frequent visits to the colony.

The black spruce aphid (*Cinara piceae*) lives in large groups on the branches of spruces, continuously feeding on their sap. During this process, the aphids produce sweet honeydew, which they secrete in large quantities. This liquid, rich in carbohydrates and water, allows ants to cover their energy and water needs.

The foragers move between the aphids and collect the drops of honeydew. Harvesting honeydew benefits both ants and aphids: ants obtain a major nutrient in large quantities, while aphids get rid of their sticky secretion. Without the ants, the aphids would eventually suffocate; fungi would cover them and damage the branches. Harvesting honeydew also stimulates the aphids' sucking activity, and this excess food increases their fertility and multiplication.
1 When a forager passes close to an aphid ready to secrete honeydew, the latter raises its abdomen and lets a droplet bead at the same time as it raises and waves its hind legs. This behavior attracts the ant's attention.

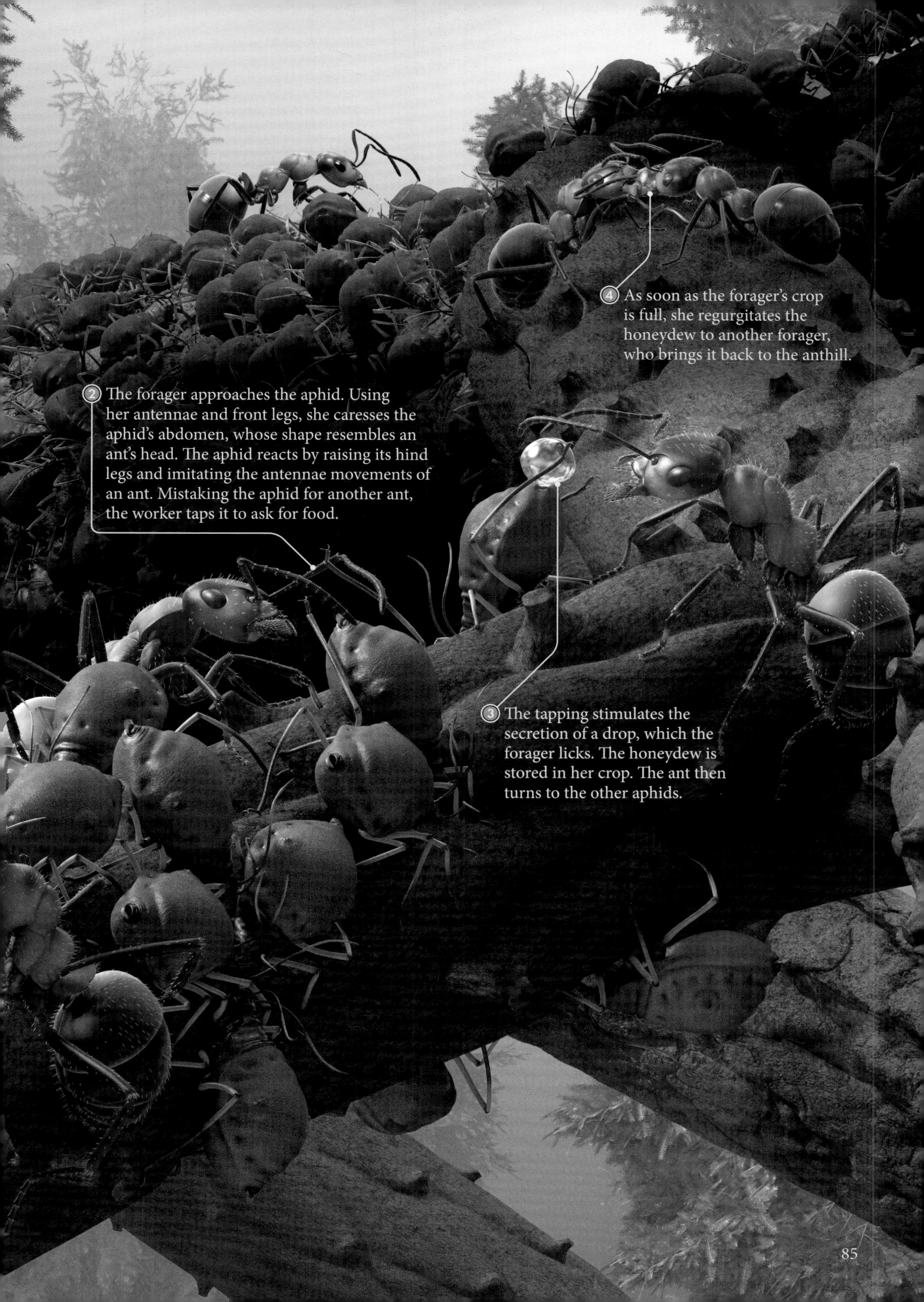
④ As soon as the forager's crop is full, she regurgitates the honeydew to another forager, who brings it back to the anthill.
② The forager approaches the aphid. Using her antennae and front legs, she caresses the aphid's abdomen, whose shape resembles an ant's head. The aphid reacts by raising its hind legs and imitating the antennae movements of an ant. Mistaking the aphid for another ant, the worker taps it to ask for food.
③ The tapping stimulates the secretion of a drop, which the forager licks. The honeydew is stored in her crop. The ant then turns to the other aphids.

6.2 Enemies and competitors

The aphids' relationship with ants provides a great advantage to the aphid colony: the ants defend the colony against both predators and competitors who would like to seize the honeydew.

The colony is constantly visited by ants. Thanks to their presence, they quickly notice predators that threaten the aphids. When attacked, they rush forward, assume a threatening pose, and strive to drive the intruder away.

When they come in contact with a predator, the aphids emit an alarm pheromone that spreads around them and causes their congeners to flee.

Aphids are ladybugs' main prey. The ladybugs are well protected from ant attacks by their thick carapace. Nevertheless, the ants manage to fend them off by using their mandibles to cling to them and throw them off the branch.

Honeybees are also attracted to honeydew, from which they make forest honey. Due to the ants' intense care, the aphid colonies are larger and produce more honeydew than colonies without ants. This more abundant honeydew interests the bees. But those who get too close to the colony are chased away by the guards.

As the peripheral groups of aphids receive less frequent visits from ants, the aphids splatter the excess honeydew around them. Unlike ants, bees do not milk the aphids but lick up the splattered drops of honeydew.

By sucking the sap, the aphids deprive the young shoots of nutrients. These wither and die. In turn, the aphids are deprived of food and stop producing honeydew. The ants then help the aphids migrate to a fresh shoot.

6.3 Plant food

Plant foods are rich in carbohydrates and provide the workers with the energy necessary for their activities. About two-thirds of their diet is plant based.

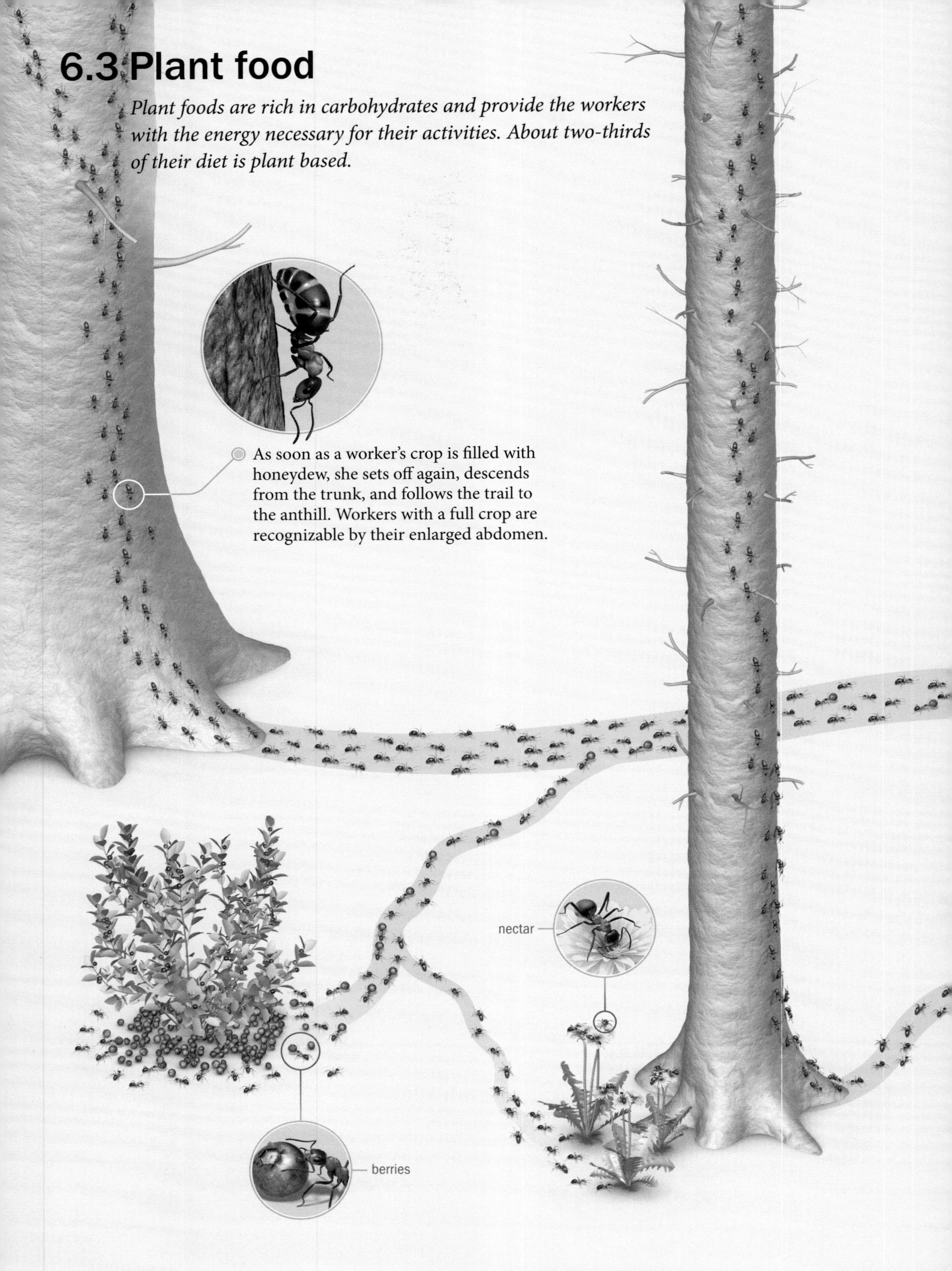

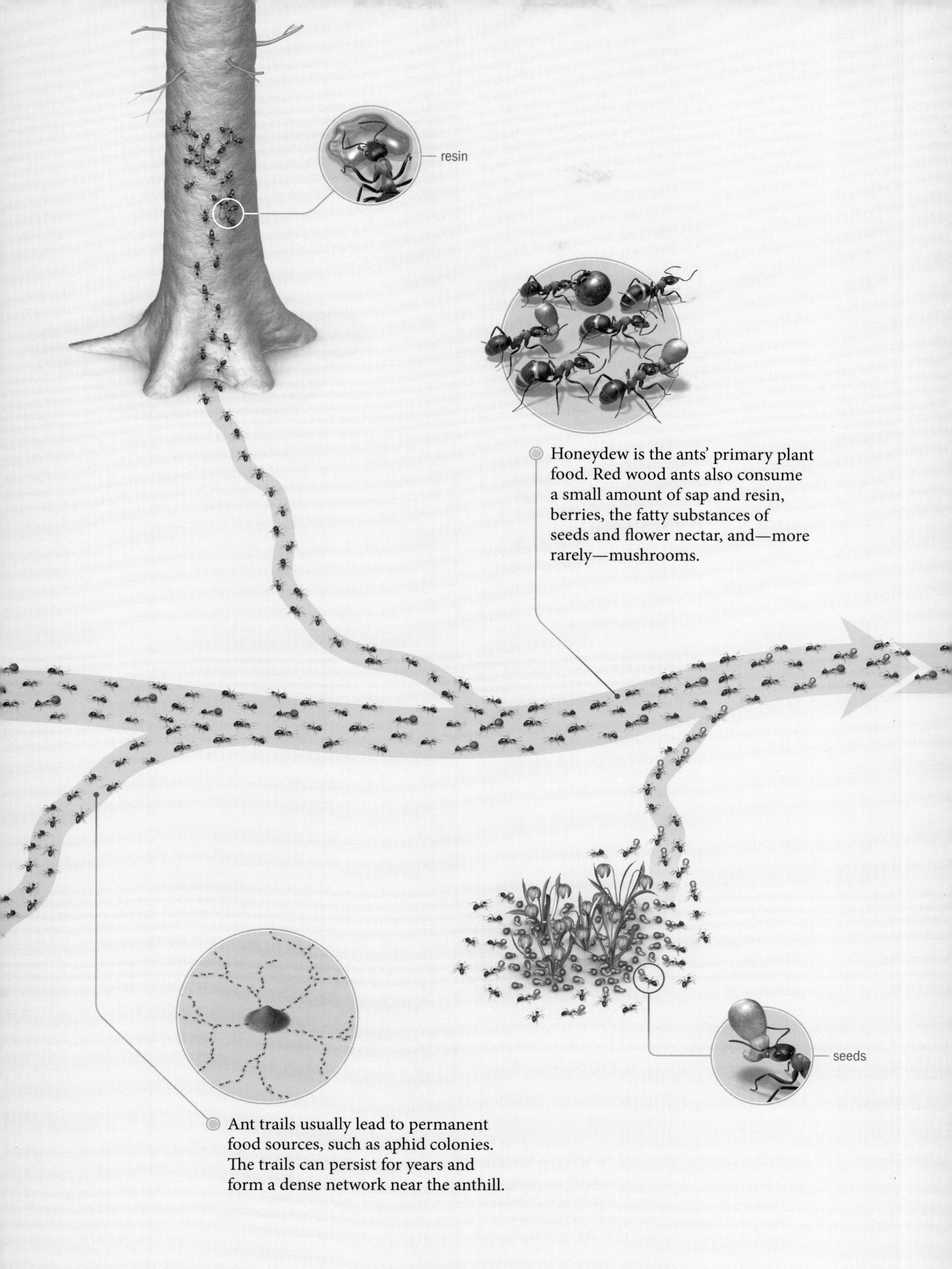

Honeydew is the ants' primary plant food. Red wood ants also consume a small amount of sap and resin, berries, the fatty substances of seeds and flower nectar, and—more rarely—mushrooms.

Ant trails usually lead to permanent food sources, such as aphid colonies. The trails can persist for years and form a dense network near the anthill.

6.4 Animal food

Animal prey is a protein source for the colony, which is essential for raising the larvae and feeding the queen. About a third of the red wood ants' food is of animal origin.

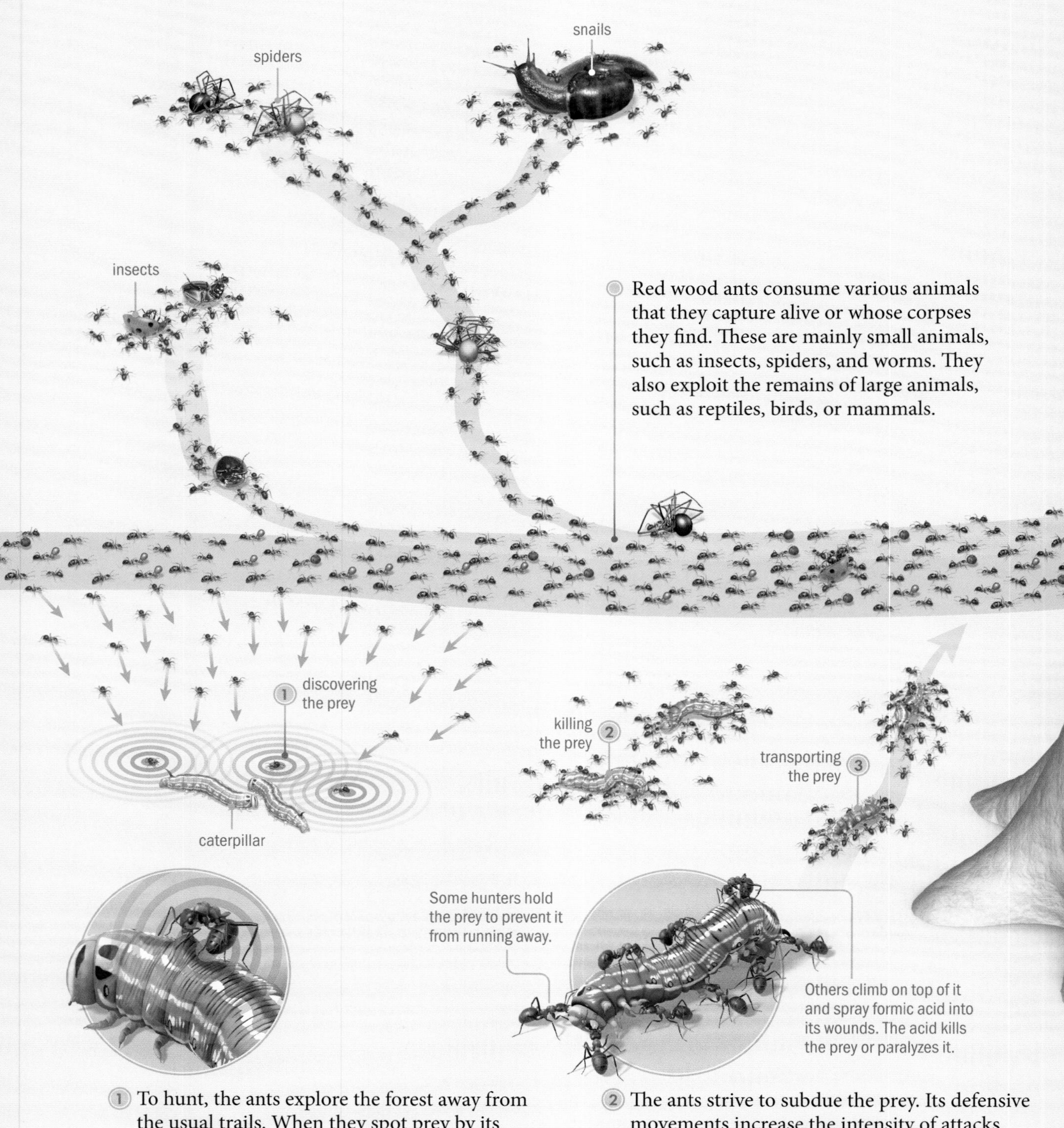

① To hunt, the ants explore the forest away from the usual trails. When they spot prey by its movements, they approach and touch it with their antennae. As soon as they have identified it as prey, they attack. The odor of the formic acid projected during the struggle and the prey's defensive movements attract other ants.

② The ants strive to subdue the prey. Its defensive movements increase the intensity of attacks. There is a division of labor among the hunters.

③ Once the prey has been subdued, the ants drag it toward the trail and from there toward the anthill, regularly taking turns transporting it.

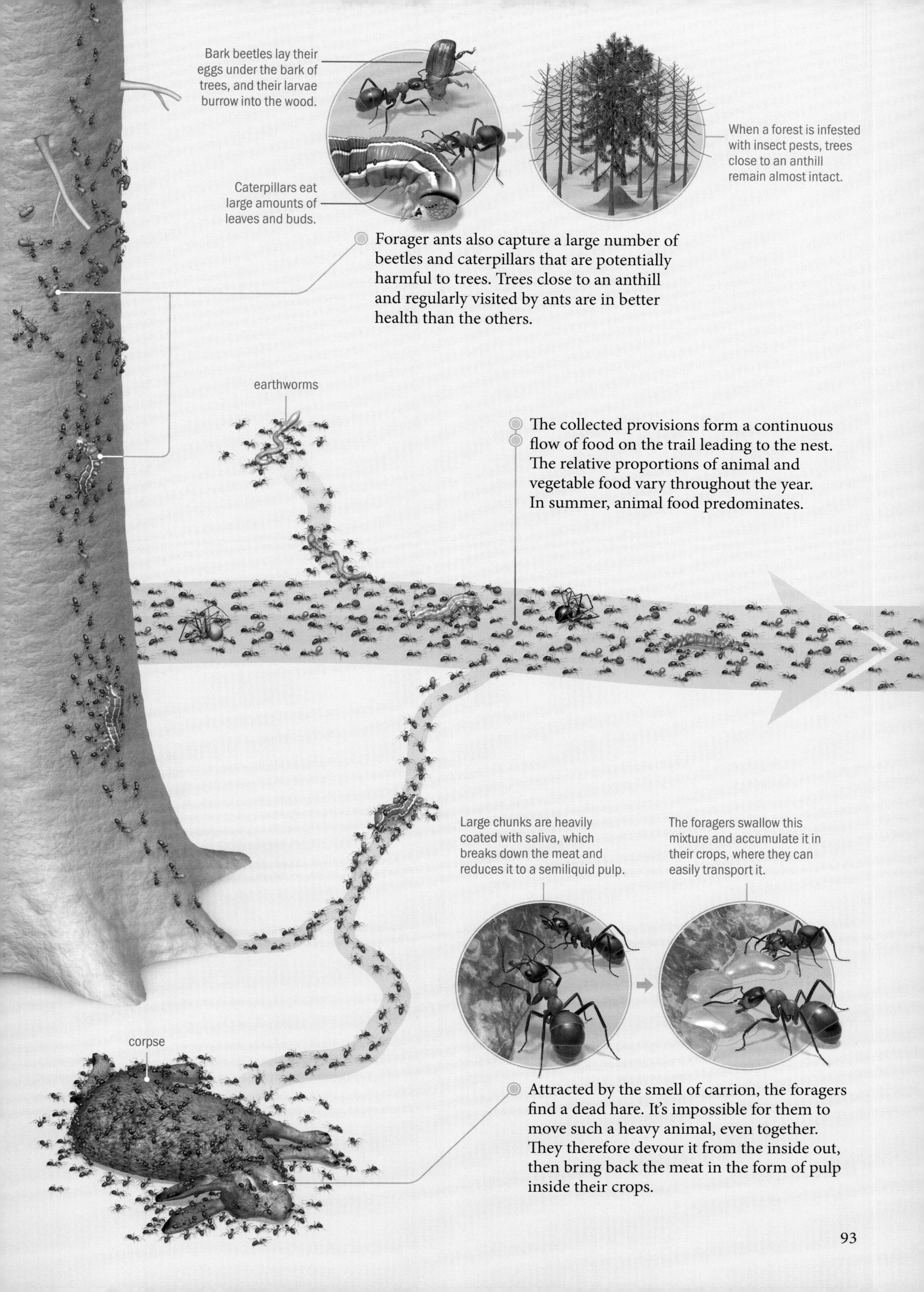

Forager ants also capture a large number of beetles and caterpillars that are potentially harmful to trees. Trees close to an anthill and regularly visited by ants are in better health than the others.

The collected provisions form a continuous flow of food on the trail leading to the nest. The relative proportions of animal and vegetable food vary throughout the year. In summer, animal food predominates.

Attracted by the smell of carrion, the foragers find a dead hare. It's impossible for them to move such a heavy animal, even together. They therefore devour it from the inside out, then bring back the meat in the form of pulp inside their crops.

6.5 Food sharing

In the anthill, food is shared between the members of the colony in different ways according to their natural requirements.

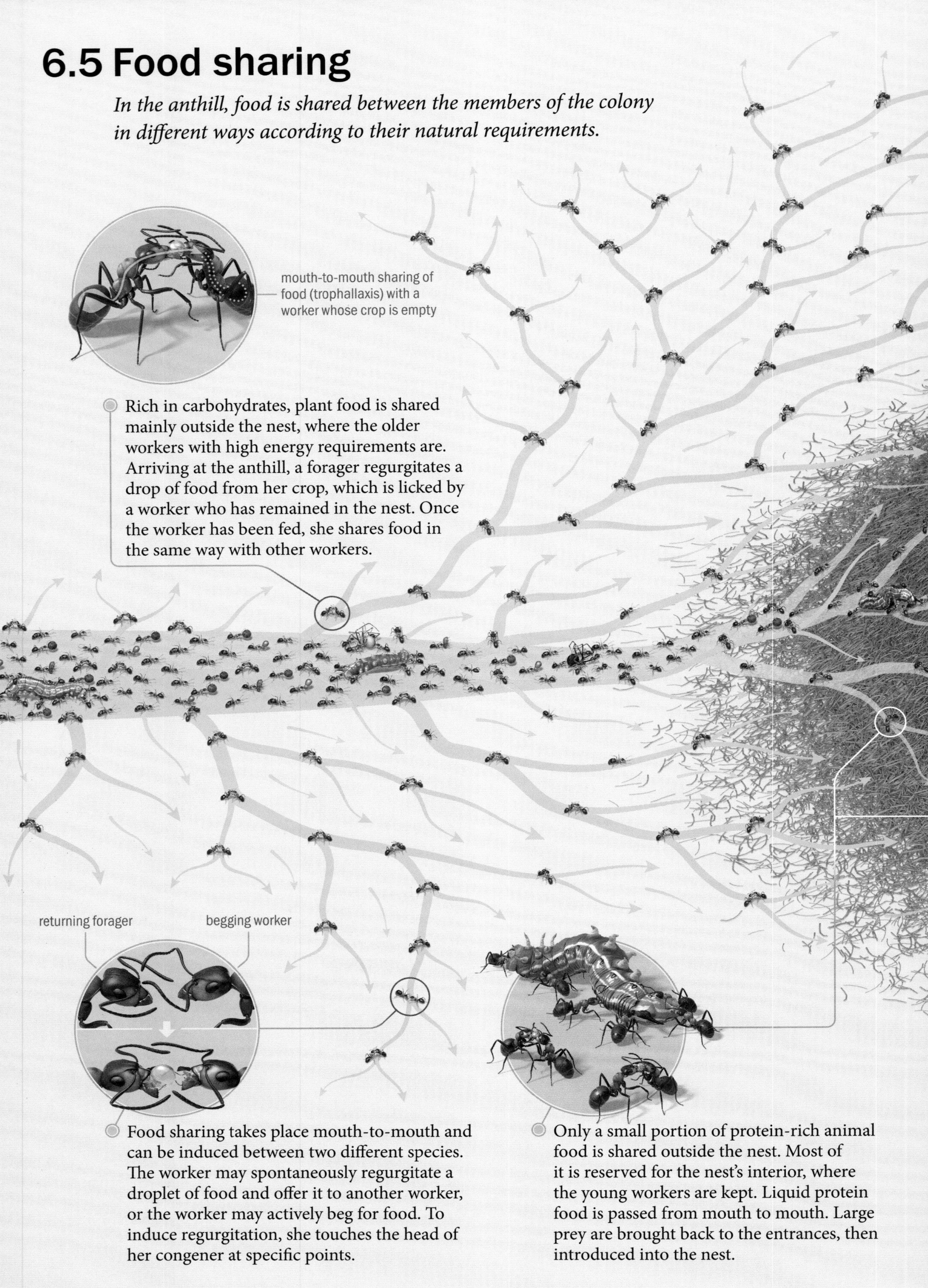

Rich in carbohydrates, plant food is shared mainly outside the nest, where the older workers with high energy requirements are. Arriving at the anthill, a forager regurgitates a drop of food from her crop, which is licked by a worker who has remained in the nest. Once the worker has been fed, she shares food in the same way with other workers.

Food sharing takes place mouth-to-mouth and can be induced between two different species. The worker may spontaneously regurgitate a droplet of food and offer it to another worker, or the worker may actively beg for food. To induce regurgitation, she touches the head of her congener at specific points.

Only a small portion of protein-rich animal food is shared outside the nest. Most of it is reserved for the nest's interior, where the young workers are kept. Liquid protein food is passed from mouth to mouth. Large prey are brought back to the entrances, then introduced into the nest.

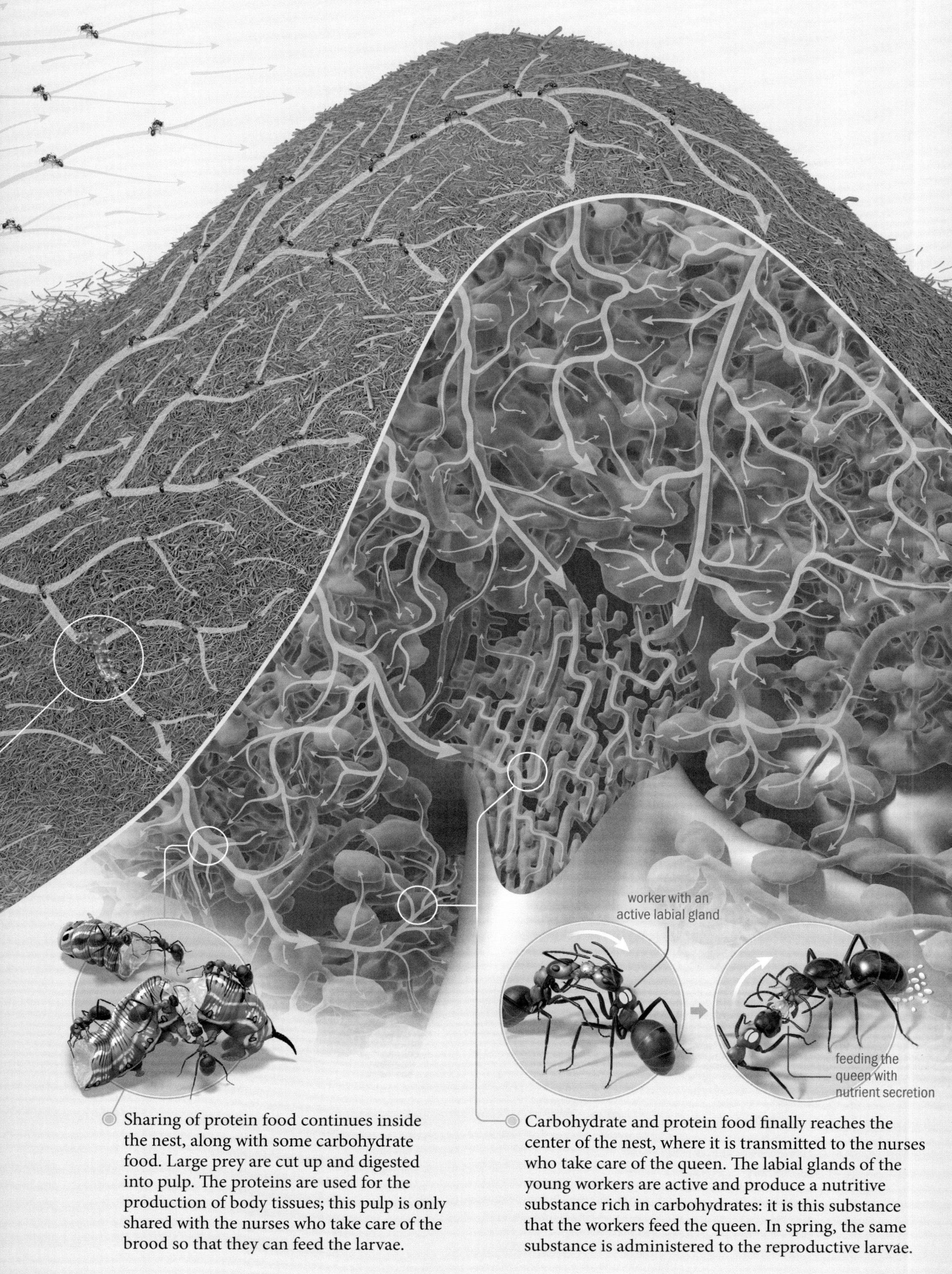

Sharing of protein food continues inside the nest, along with some carbohydrate food. Large prey are cut up and digested into pulp. The proteins are used for the production of body tissues; this pulp is only shared with the nurses who take care of the brood so that they can feed the larvae.

Carbohydrate and protein food finally reaches the center of the nest, where it is transmitted to the nurses who take care of the queen. The labial glands of the young workers are active and produce a nutritive substance rich in carbohydrates: it is this substance that the workers feed the queen. In spring, the same substance is administered to the reproductive larvae.

6.6 Waste elimination

Maintaining nest hygiene is an important task among ants. Because the heat and humidity that prevail inside promote the development of pathogenic microbes and fungi, all potentially dangerous substances are evacuated from the anthill.

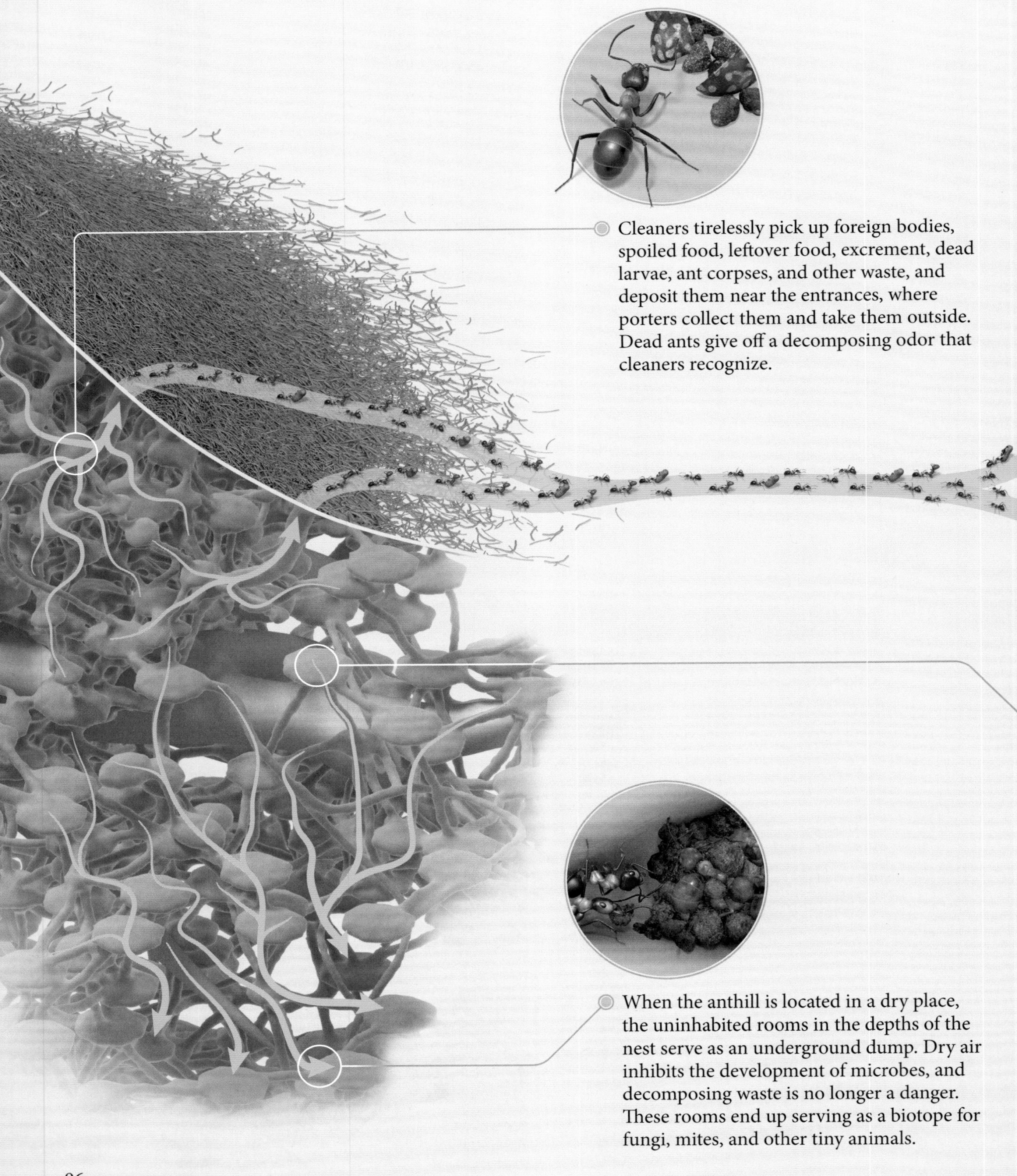

Cleaners tirelessly pick up foreign bodies, spoiled food, leftover food, excrement, dead larvae, ant corpses, and other waste, and deposit them near the entrances, where porters collect them and take them outside. Dead ants give off a decomposing odor that cleaners recognize.

When the anthill is located in a dry place, the uninhabited rooms in the depths of the nest serve as an underground dump. Dry air inhibits the development of microbes, and decomposing waste is no longer a danger. These rooms end up serving as a biotope for fungi, mites, and other tiny animals.

When the depths of the nest are unsuitable for waste storage, the waste is carried out of the nest and deposited in heaps, generally in dry places, just on the outskirts of the colony's hunting territory. These heaps of waste are used to mark the territory limits.

A large number of tiny insects contribute to the nest's hygiene by feeding on vegetable matter, leftover food, and the corpses of decomposing ants. These include springtails, silverfish, beetle larvae (white grubs), or beetles of the genus *Dinarda*, with atrophied wings. Some of these commensals are tolerated or watched over by the workers. Others are difficult to seize because of their flattened bodies or defensive substances.

7.1 Origin of sex and caste

It is during larval development that the ant's sex and caste is determined. This development is influenced by the queen and the workers.

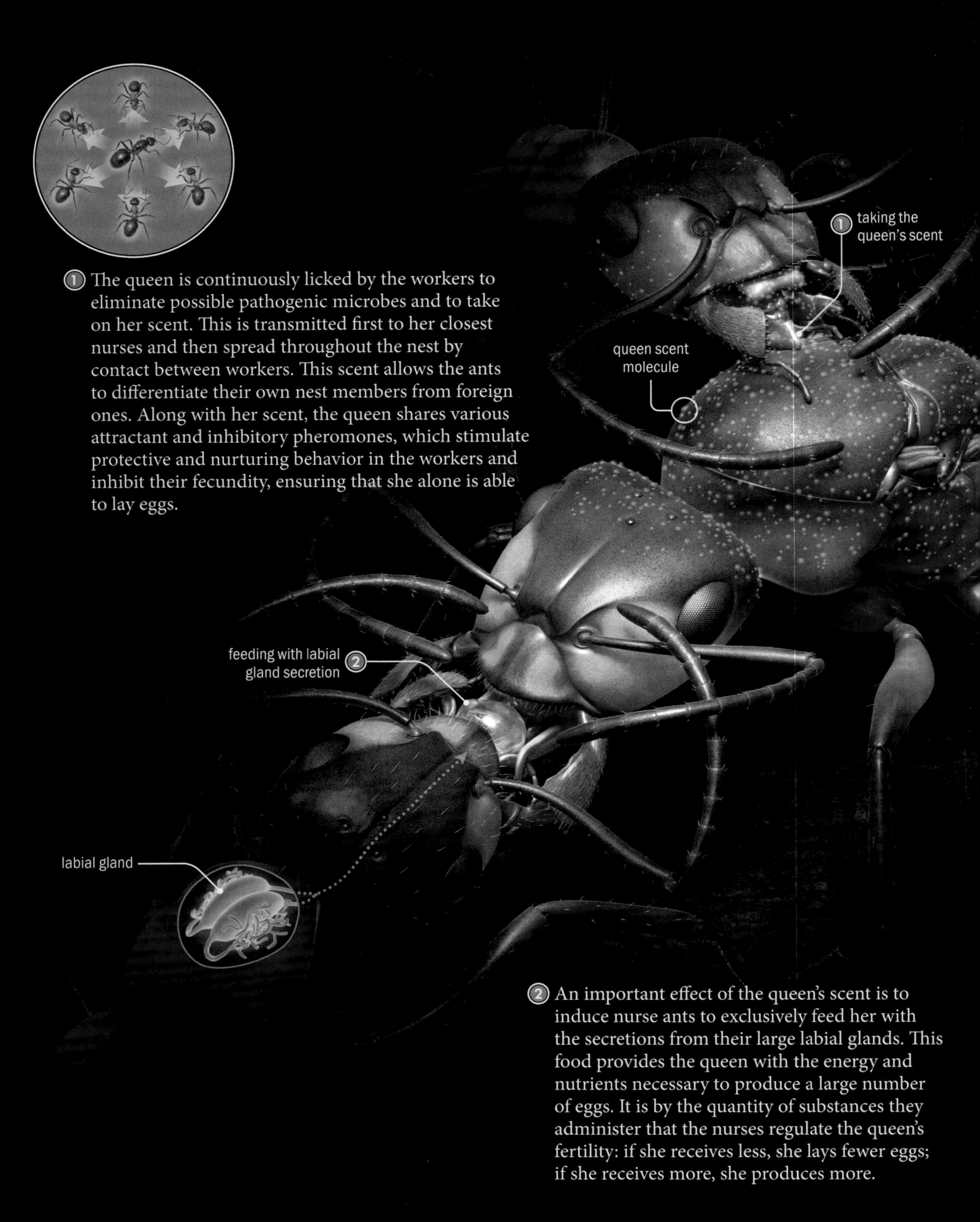

① The queen is continuously licked by the workers to eliminate possible pathogenic microbes and to take on her scent. This is transmitted first to her closest nurses and then spread throughout the nest by contact between workers. This scent allows the ants to differentiate their own nest members from foreign ones. Along with her scent, the queen shares various attractant and inhibitory pheromones, which stimulate protective and nurturing behavior in the workers and inhibit their fecundity, ensuring that she alone is able to lay eggs.

② An important effect of the queen's scent is to induce nurse ants to exclusively feed her with the secretions from their large labial glands. This food provides the queen with the energy and nutrients necessary to produce a large number of eggs. It is by the quantity of substances they administer that the nurses regulate the queen's fertility: if she receives less, she lays fewer eggs; if she receives more, she produces more.

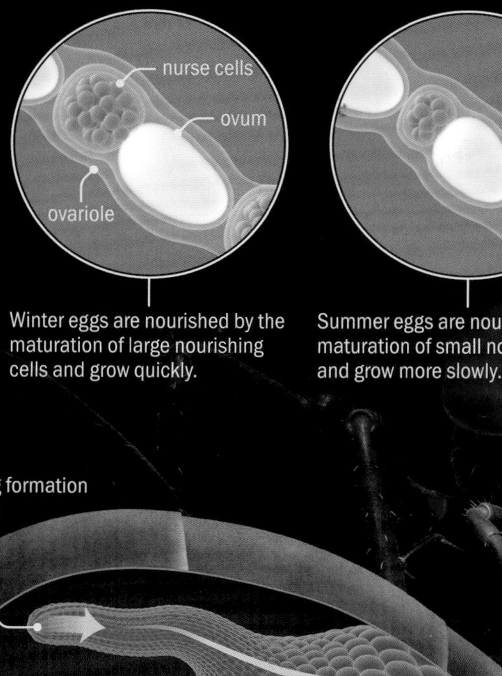

Winter eggs are nourished by the maturation of large nourishing cells and grow quickly.

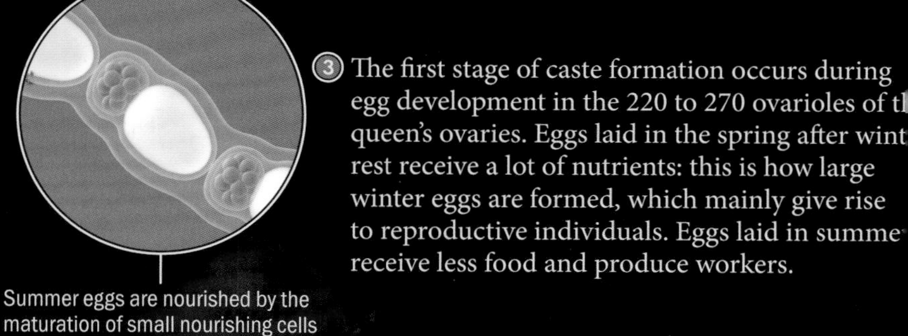

Summer eggs are nourished by the maturation of small nourishing cells and grow more slowly.

③ The first stage of caste formation occurs during egg development in the 220 to 270 ovarioles of th queen's ovaries. Eggs laid in the spring after wint rest receive a lot of nutrients: this is how large winter eggs are formed, which mainly give rise to reproductive individuals. Eggs laid in summe receive less food and produce workers.

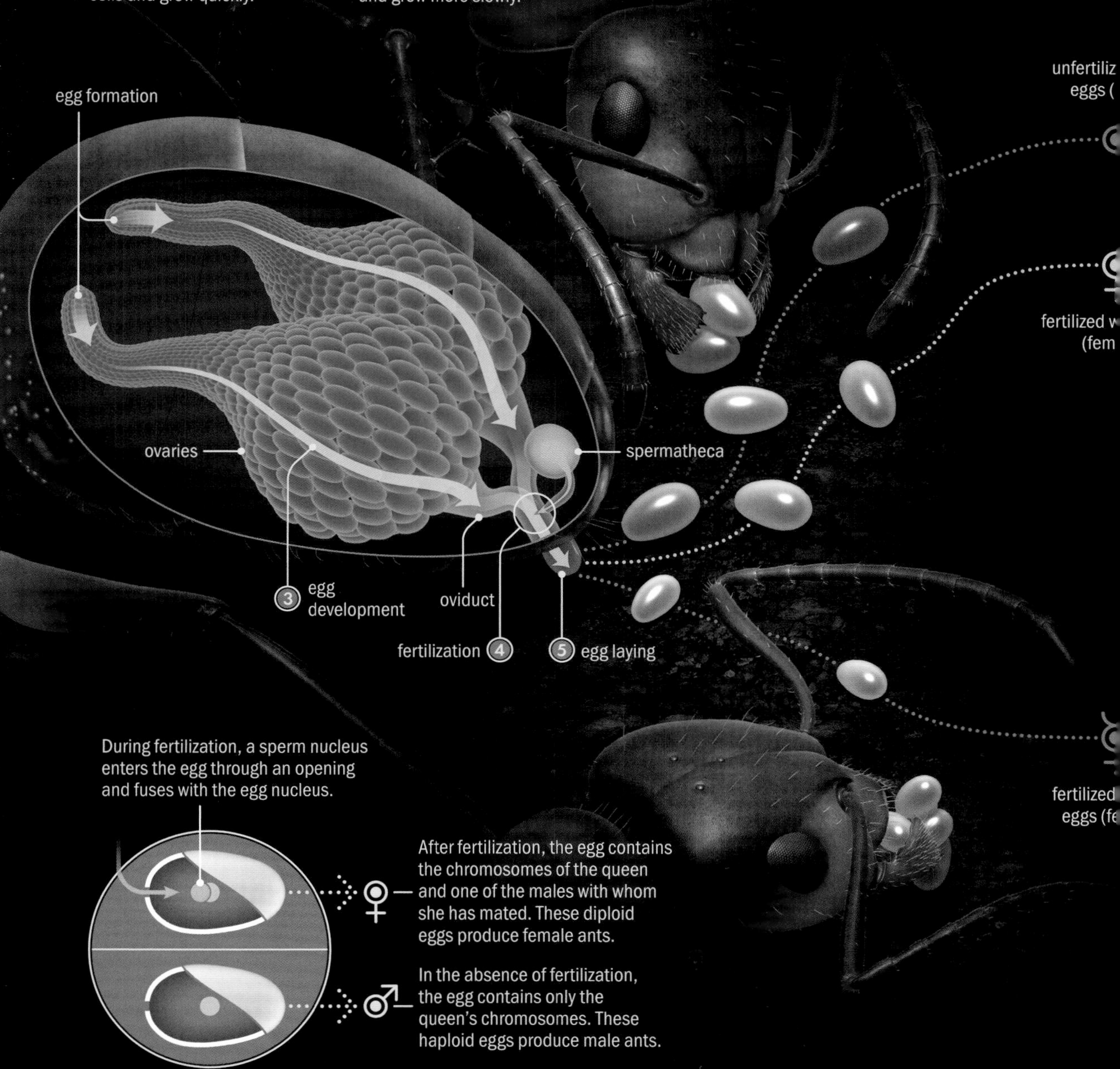

During fertilization, a sperm nucleus enters the egg through an opening and fuses with the egg nucleus.

After fertilization, the egg contains the chromosomes of the queen and one of the males with whom she has mated. These diploid eggs produce female ants.

In the absence of fertilization, the egg contains only the queen's chromosomes. These haploid eggs produce male ants.

④ The developed eggs pass into the oviduct, where they can be fertilized by sperm and expelled from the spermatheca by a muscle. This process is temperature sensitive and only works above 20°C. Thus, depending on the temperature variations in the spring, the winter eggs may or may not be fertilized. Summer eggs, on the other hand, are almost always fertilized. The individual's sex is

⑤ The oviduct opens to the exterior at the tip of the abdomen. As soon as an egg appears, it is seized by a worker and carried to an incubatin chamber. The queen lays up to 300 eggs a day for 5–6 months per year

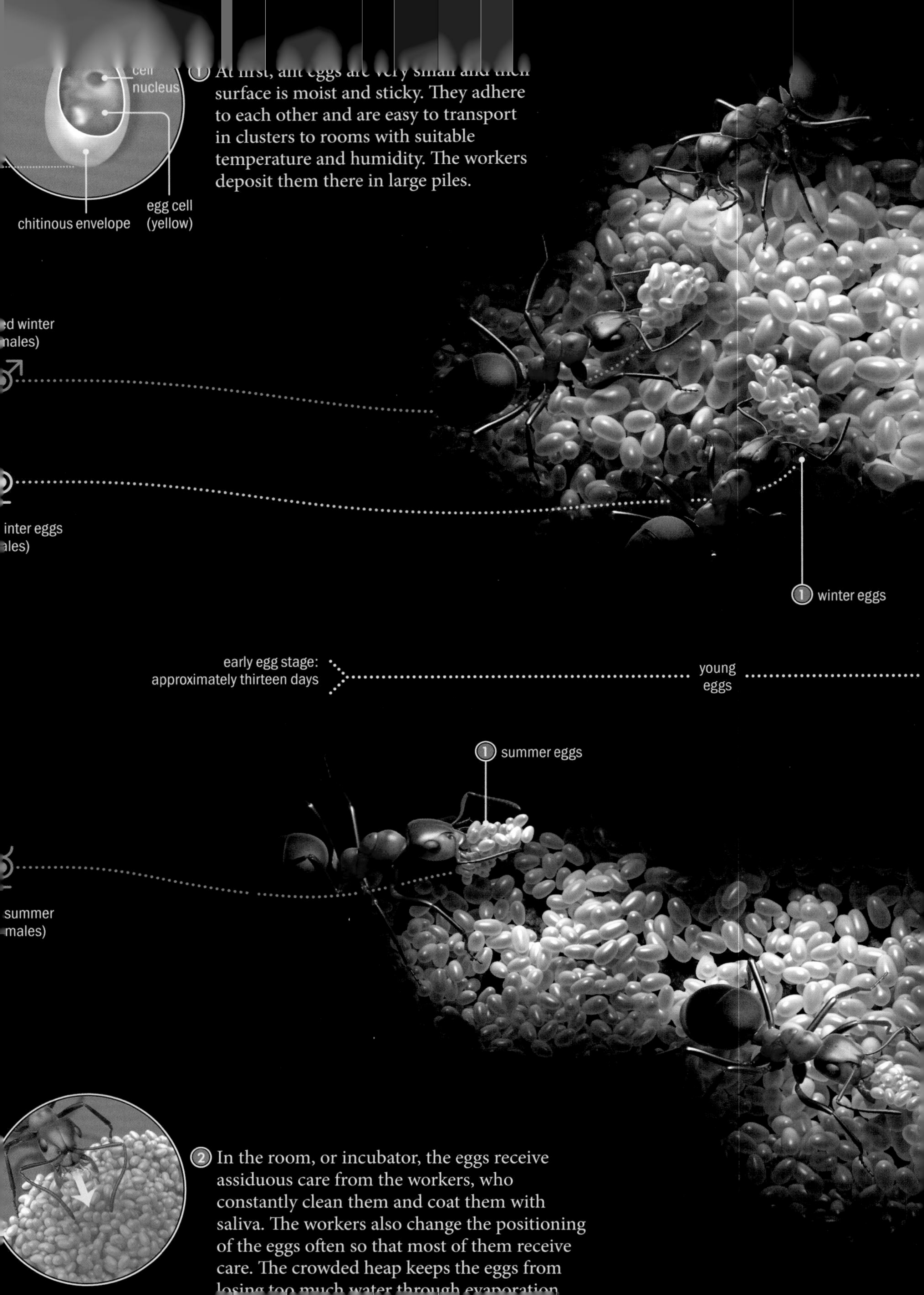
cell
nucleus
egg cell
(yellow)
chitinous envelope
① At first, ant eggs are very small and their surface is moist and sticky. They adhere to each other and are easy to transport in clusters to rooms with suitable temperature and humidity. The workers deposit them there in large piles.
ed winter
males)
inter eggs
ales)
① winter eggs
early egg stage:
approximately thirteen days
young
eggs
① summer eggs
summer
males)
② In the room, or incubator, the eggs receive assiduous care from the workers, who constantly clean them and coat them with saliva. The workers also change the positioning of the eggs often so that most of them receive care. The crowded heap keeps the eggs from losing too much water through evaporation.

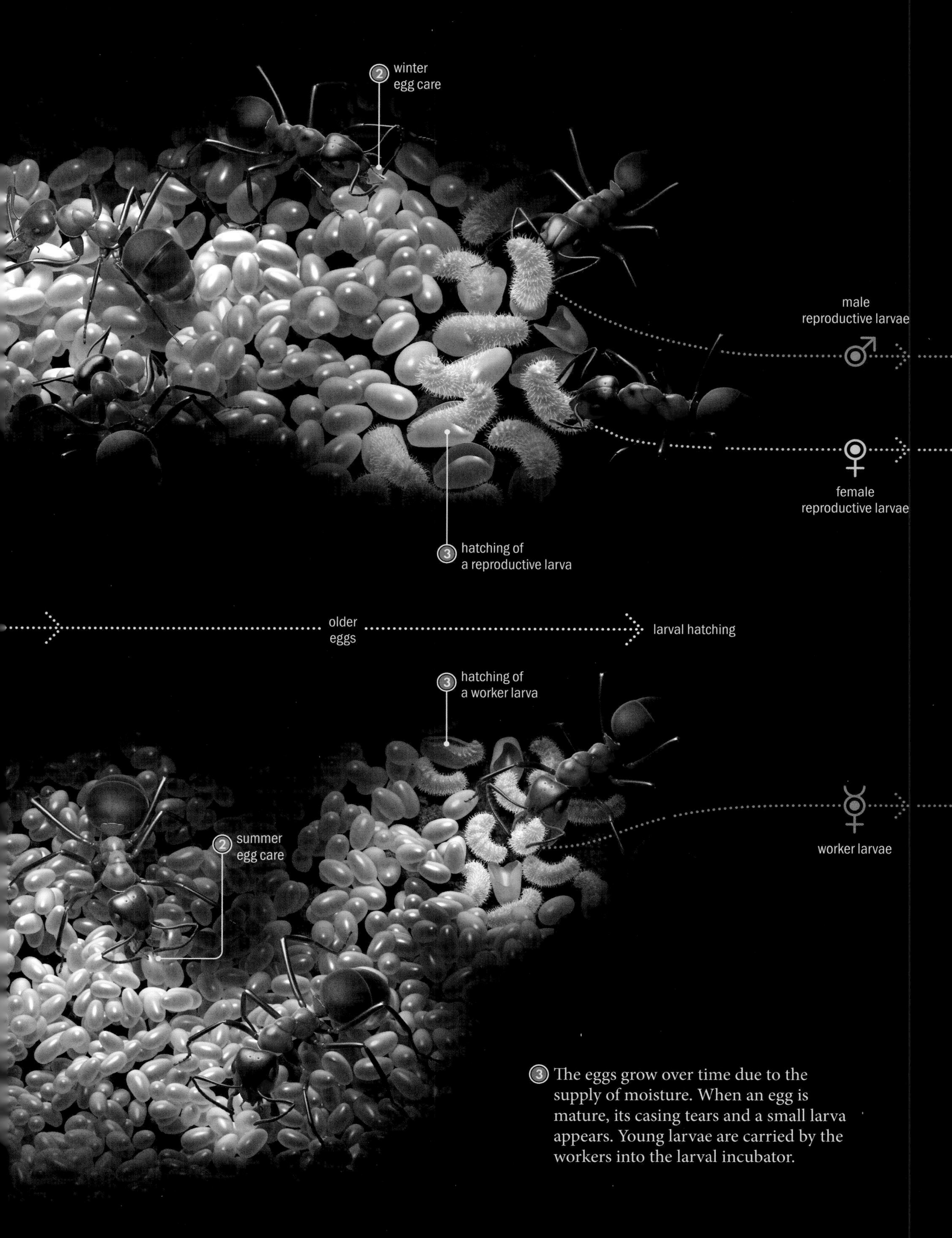

③ The eggs grow over time due to the supply of moisture. When an egg is mature, its casing tears and a small larva appears. Young larvae are carried by the workers into the larval incubator.

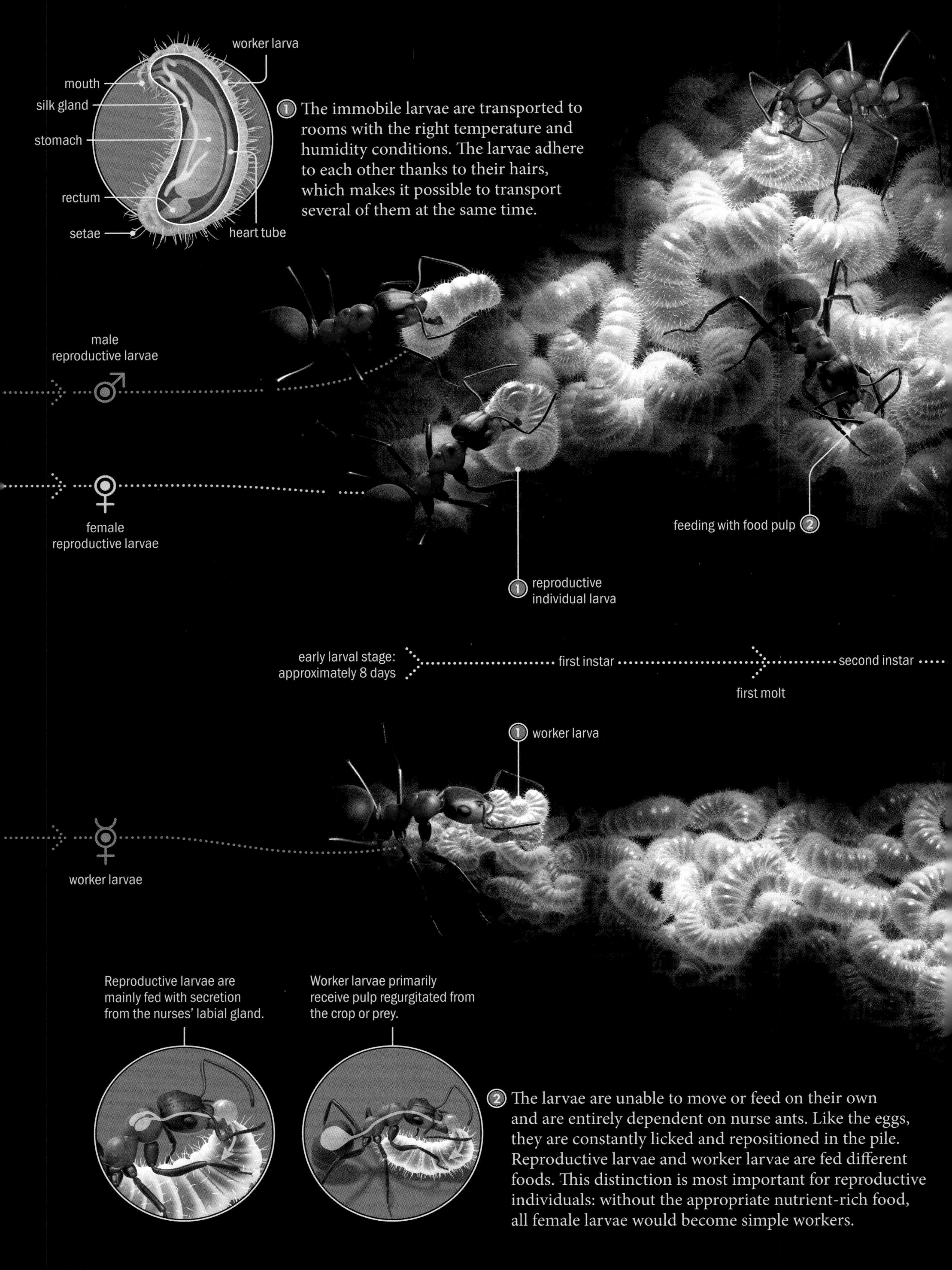

① The immobile larvae are transported to rooms with the right temperature and humidity conditions. The larvae adhere to each other thanks to their hairs, which makes it possible to transport several of them at the same time.

② The larvae are unable to move or feed on their own and are entirely dependent on nurse ants. Like the eggs, they are constantly licked and repositioned in the pile. Reproductive larvae and worker larvae are fed different foods. This distinction is most important for reproductive individuals: without the appropriate nutrient-rich food, all female larvae would become simple workers.

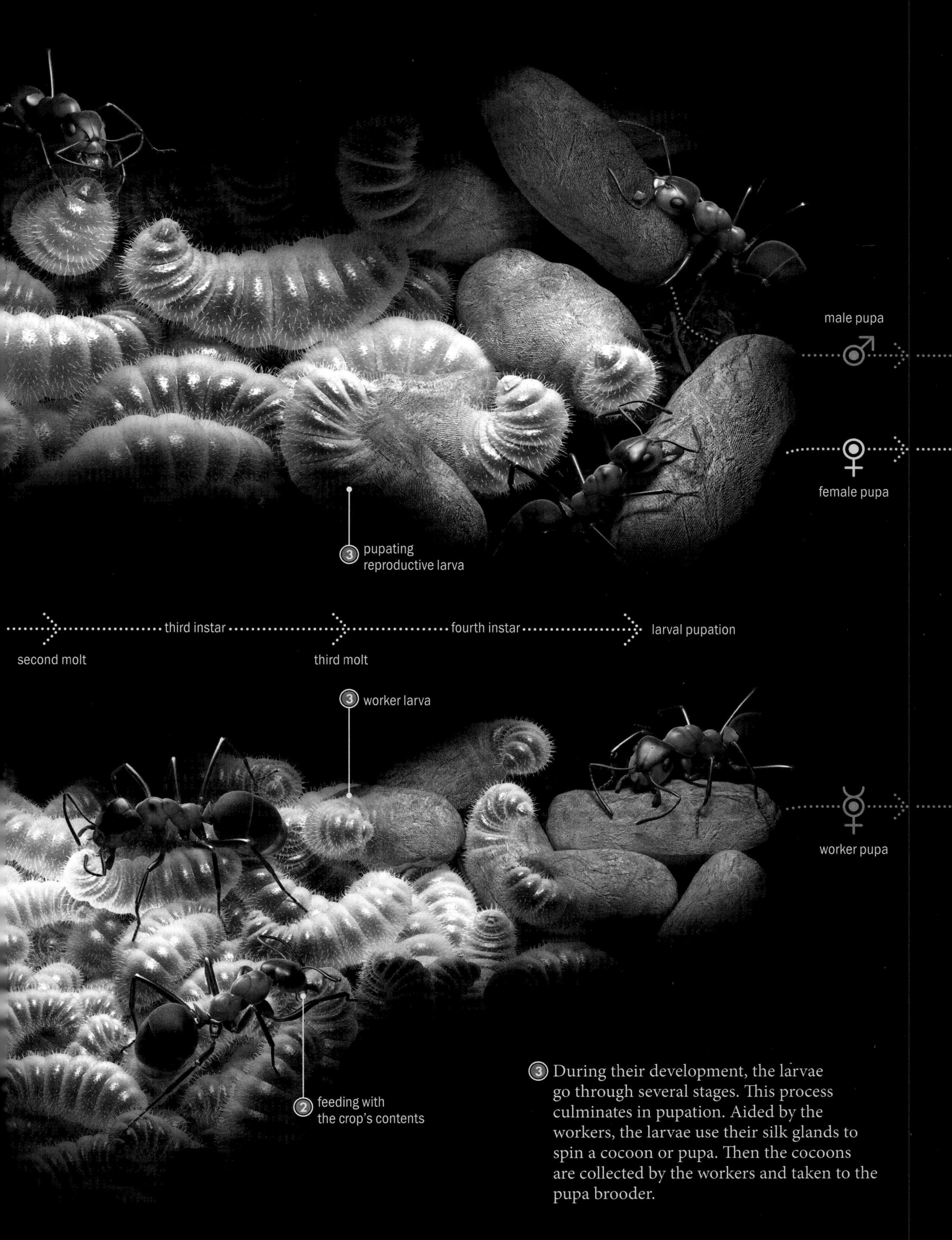

③ During their development, the larvae go through several stages. This process culminates in pupation. Aided by the workers, the larvae use their silk glands to spin a cocoon or pupa. Then the cocoons are collected by the workers and taken to the pupa brooder.

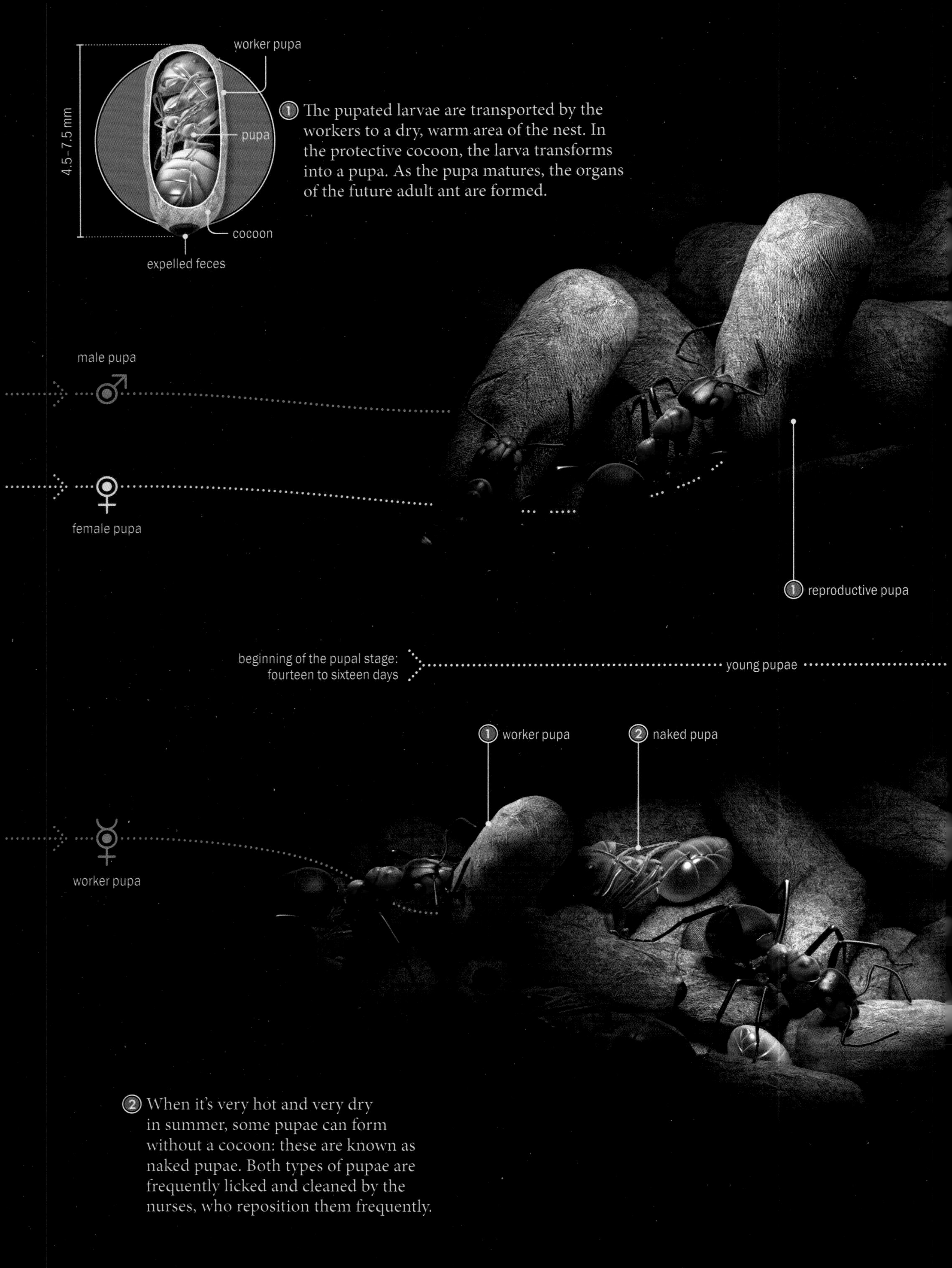

① The pupated larvae are transported by the workers to a dry, warm area of the nest. In the protective cocoon, the larva transforms into a pupa. As the pupa matures, the organs of the future adult ant are formed.

② When it's very hot and very dry in summer, some pupae can form without a cocoon: these are known as naked pupae. Both types of pupae are frequently licked and cleaned by the nurses, who reposition them frequently.

③ At the end of their development, the pupae are ready to hatch and seek to free themselves from the cocoon. The nurses help them by tearing the very thin, paper-like material.

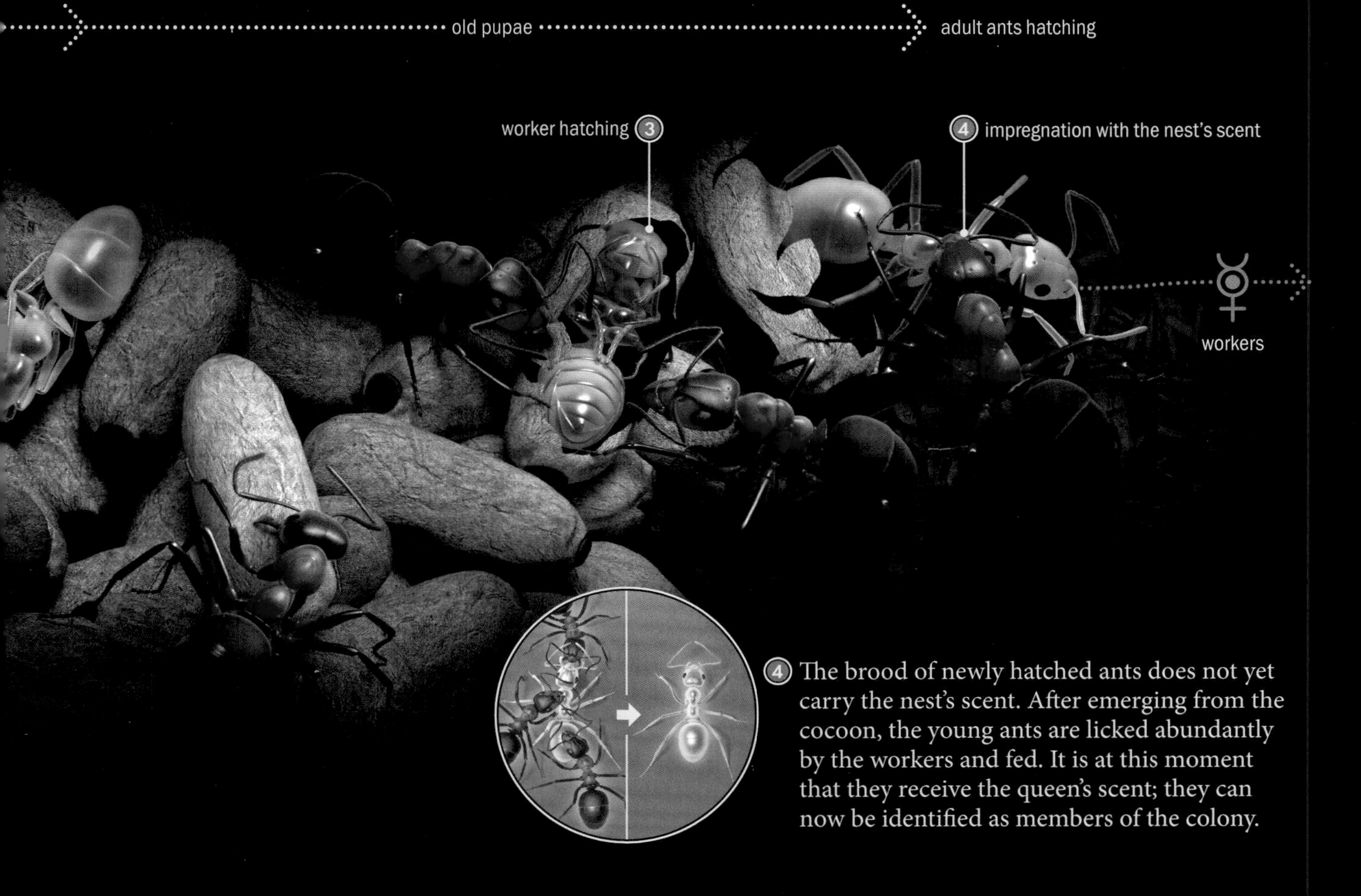

④ The brood of newly hatched ants does not yet carry the nest's scent. After emerging from the cocoon, the young ants are licked abundantly by the workers and fed. It is at this moment that they receive the queen's scent; they can now be identified as members of the colony.

7.2 The life of a reproductive individual

The vital task of male and female individuals is the survival and multiplication of the colony. The lives of both sexes can be divided into three phases.

① The first phase of the life of the newly hatched males takes place inside the nest and lasts three or four days, during which they participate in social life. They stand in the darkness of the nest and seek contact with each other. Their sex drive is still weak.

① Newly hatched females also spend several days inside the nest, taking part in social activities, like trophallaxis and mutual cleansing.

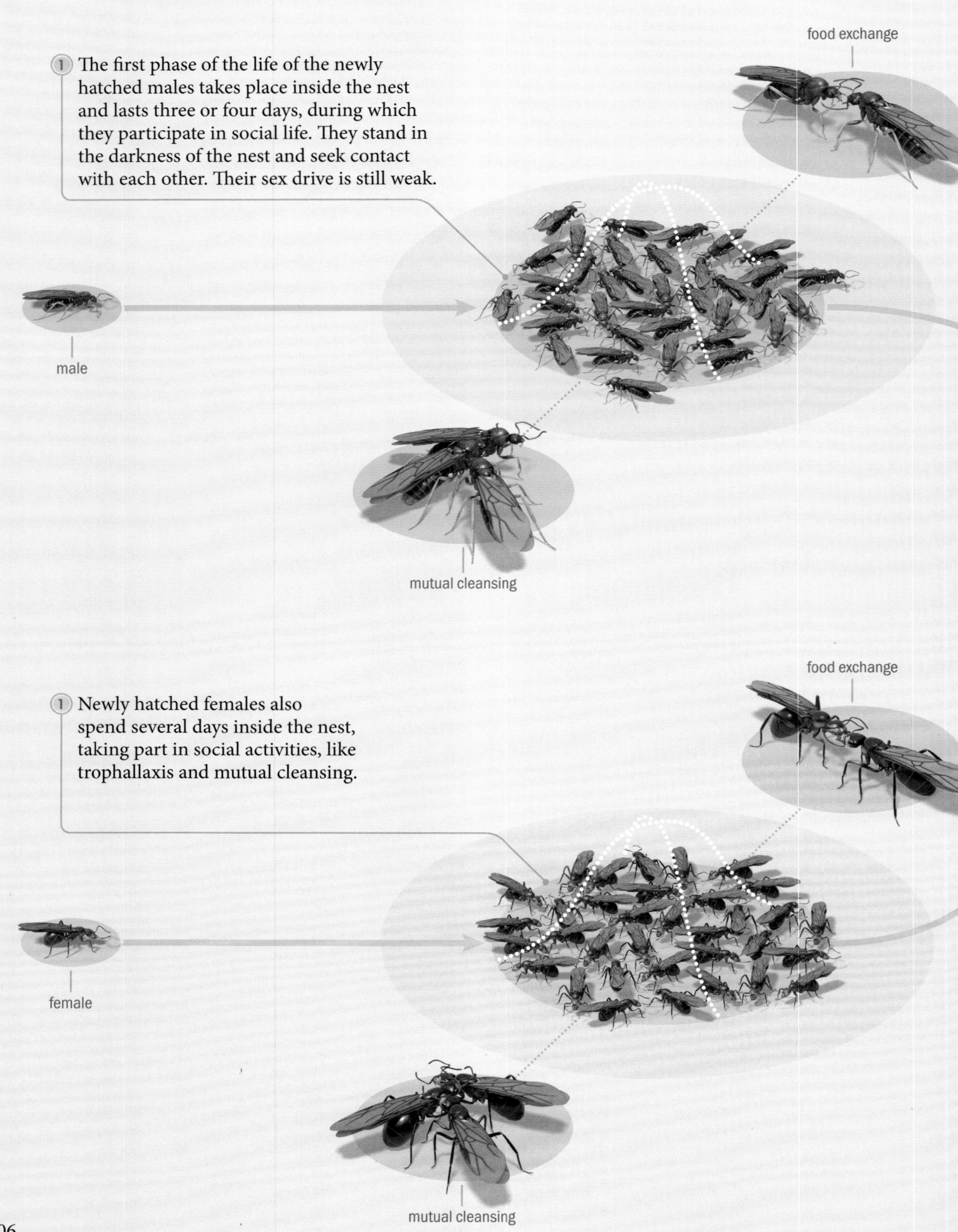

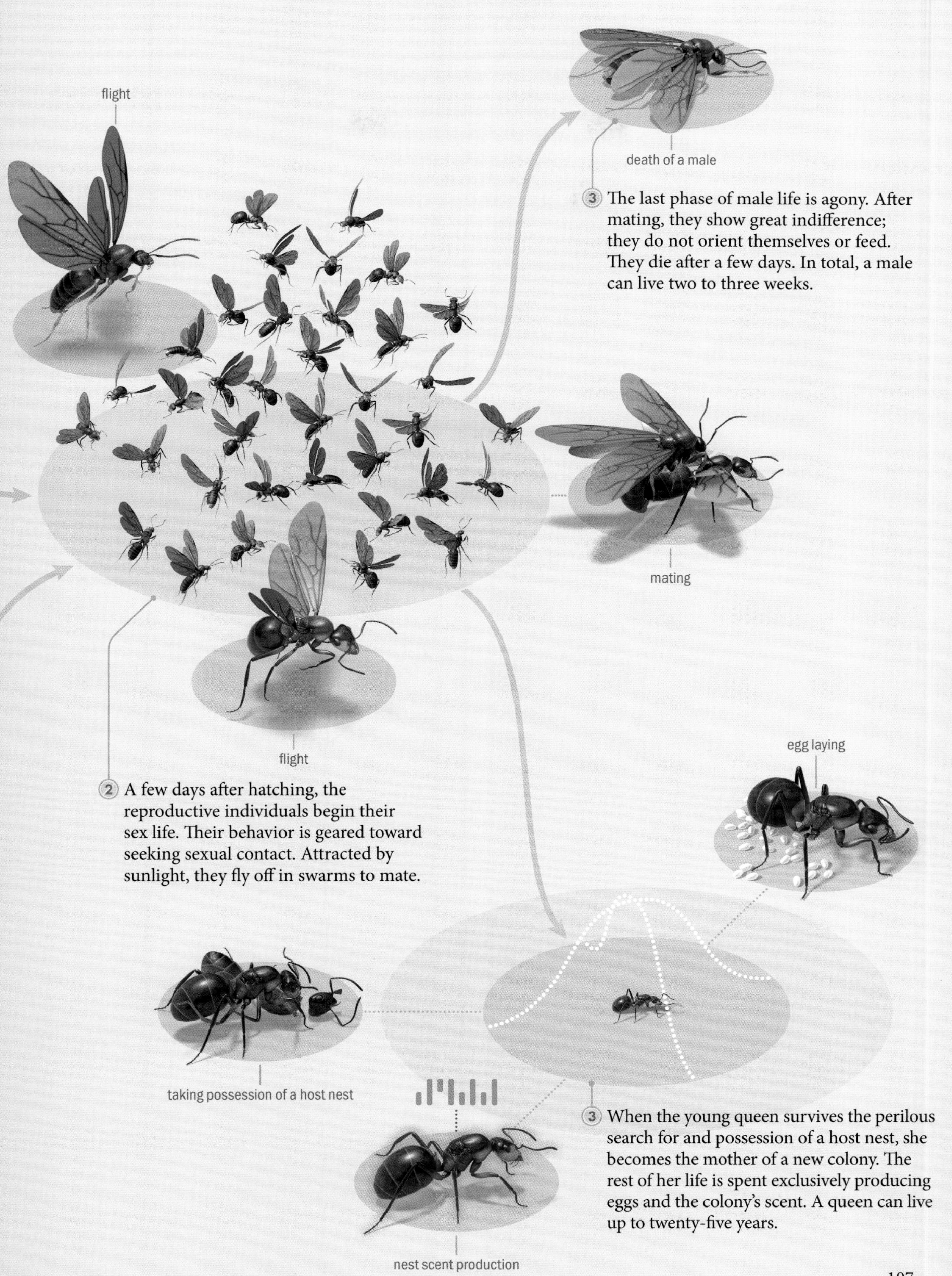

③ The last phase of male life is agony. After mating, they show great indifference; they do not orient themselves or feed. They die after a few days. In total, a male can live two to three weeks.

② A few days after hatching, the reproductive individuals begin their sex life. Their behavior is geared toward seeking sexual contact. Attracted by sunlight, they fly off in swarms to mate.

③ When the young queen survives the perilous search for and possession of a host nest, she becomes the mother of a new colony. The rest of her life is spent exclusively producing eggs and the colony's scent. A queen can live up to twenty-five years.

7.3 The life of a worker

Workers perform all the tasks necessary for the survival of the colony. During their lives, they pass through a phase of interior service and a phase of exterior service, each comprising various fields of activity.

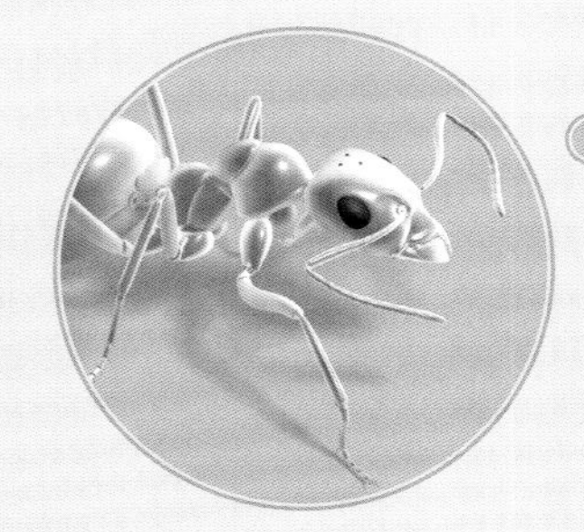

① At first, the body of a newly hatched worker is colorless and soft. In a few hours or days, its cuticle covering hardens and takes on a characteristic reddish-brown-black color. The limbs can then move and the worker begins her activity in the anthill.

② The first phase of a worker's life is devoted to interior service. She almost never leaves the nest and devotes herself to various tasks, such as brood care, cleaning, or maintenance of the rooms. An interior service ant is supplied provisions by the older exterior service workers. She only comes out of the nest on exceptional occasions, such as a food shortage or the colony's relocation.

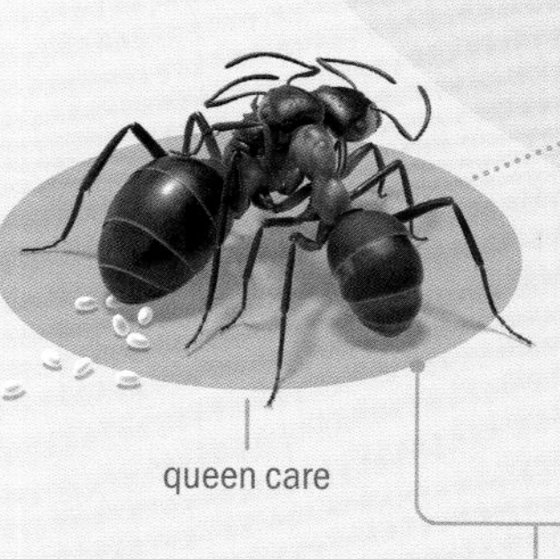

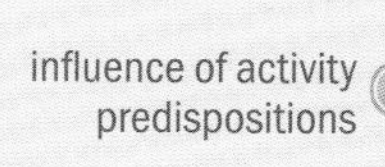

③ Activities in the nest are triggered by an olfactory stimulus to which the workers respond by busying themselves with their tasks. For them to perceive the stimulus, it must reach a certain intensity. The sensitivity of workers to these stimuli is variable and differs from one to another. Some react more to the stimulus of the queen, others to that of the eggs or larvae. This is how the workers share the different tasks inside the nest.

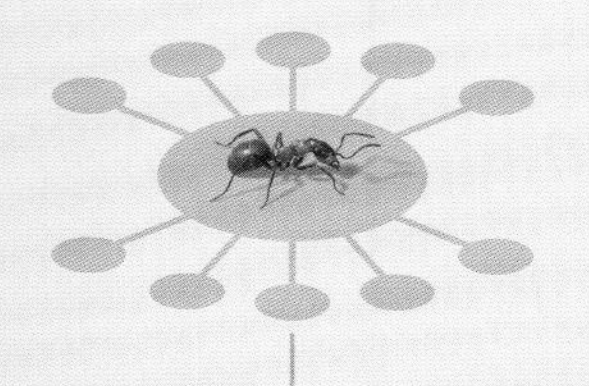

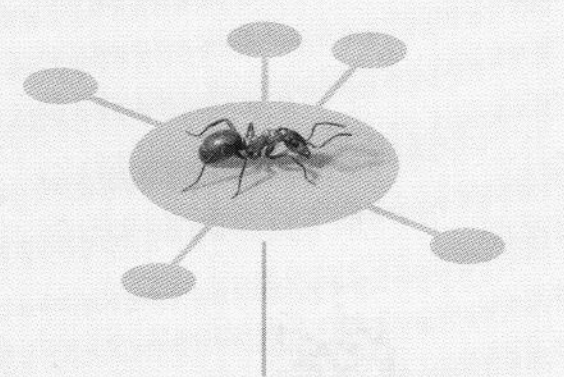

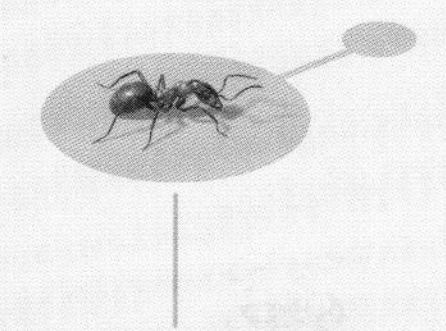

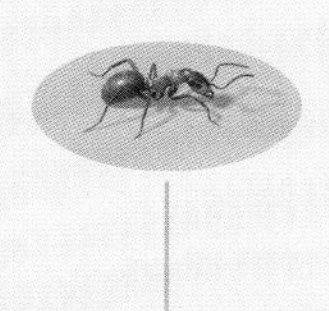

(4) Task division results not only from workers' predispositions for certain stimuli but also from the intensity of the corresponding tasks. This is why some workers may remain faithful to an activity, while others may change what they do. But a large number of workers do not devote themselves to any specific task. These passive individuals may constitute a labor reserve. When the nest is damaged, they can be mobilized without the other workers needing to neglect their duties.

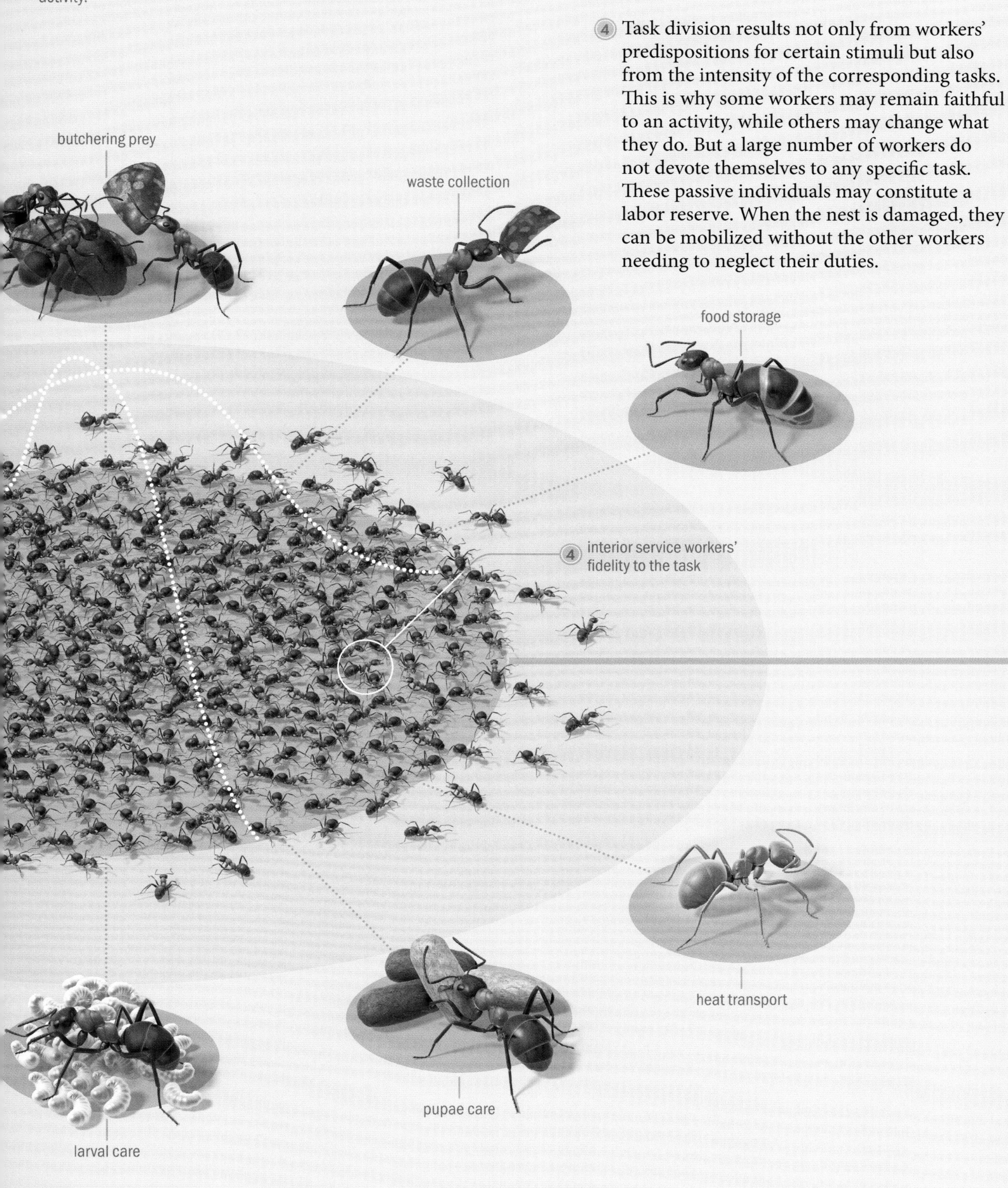

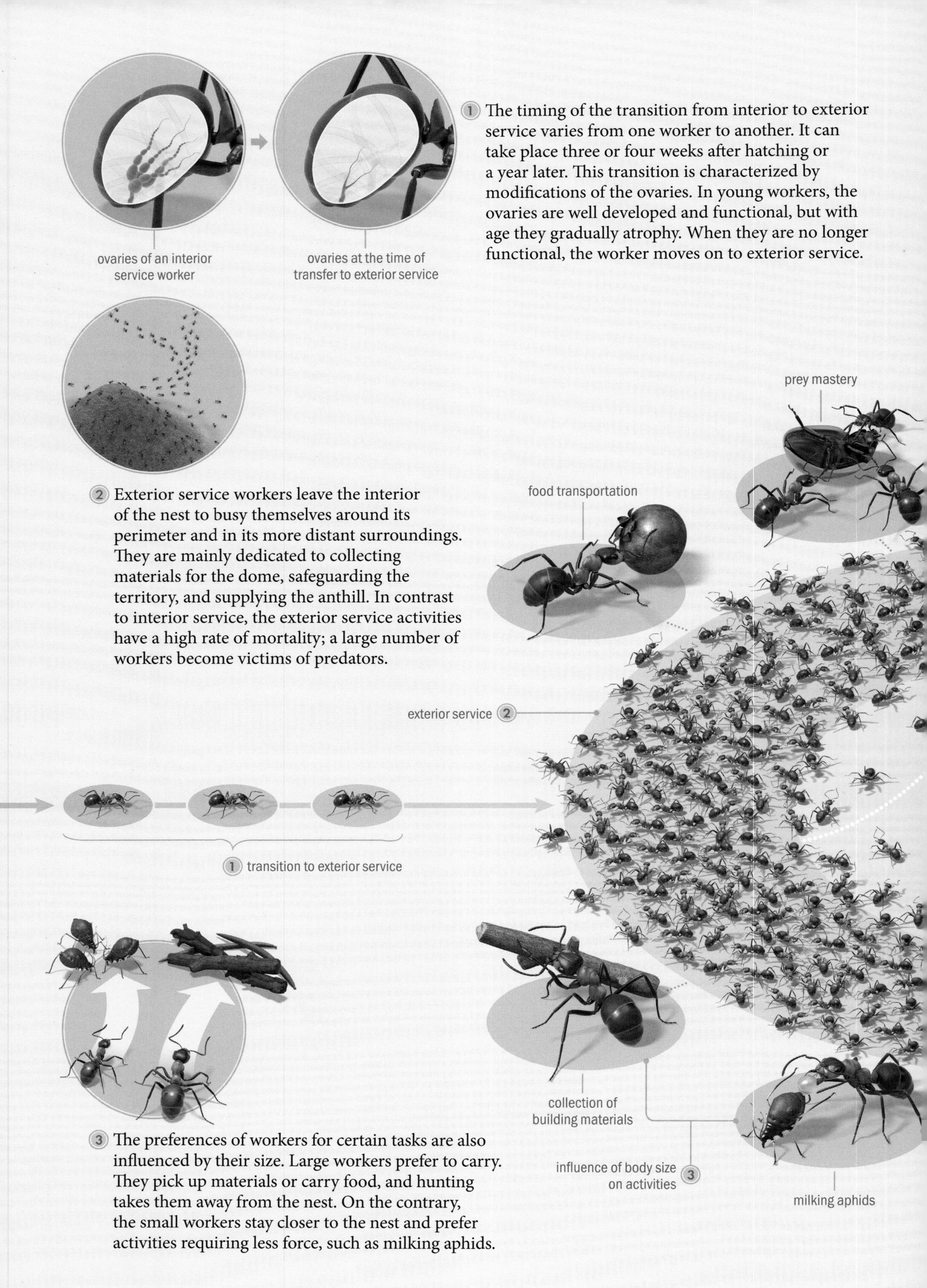

① The timing of the transition from interior to exterior service varies from one worker to another. It can take place three or four weeks after hatching or a year later. This transition is characterized by modifications of the ovaries. In young workers, the ovaries are well developed and functional, but with age they gradually atrophy. When they are no longer functional, the worker moves on to exterior service.

② Exterior service workers leave the interior of the nest to busy themselves around its perimeter and in its more distant surroundings. They are mainly dedicated to collecting materials for the dome, safeguarding the territory, and supplying the anthill. In contrast to interior service, the exterior service activities have a high rate of mortality; a large number of workers become victims of predators.

③ The preferences of workers for certain tasks are also influenced by their size. Large workers prefer to carry. They pick up materials or carry food, and hunting takes them away from the nest. On the contrary, the small workers stay closer to the nest and prefer activities requiring less force, such as milking aphids.

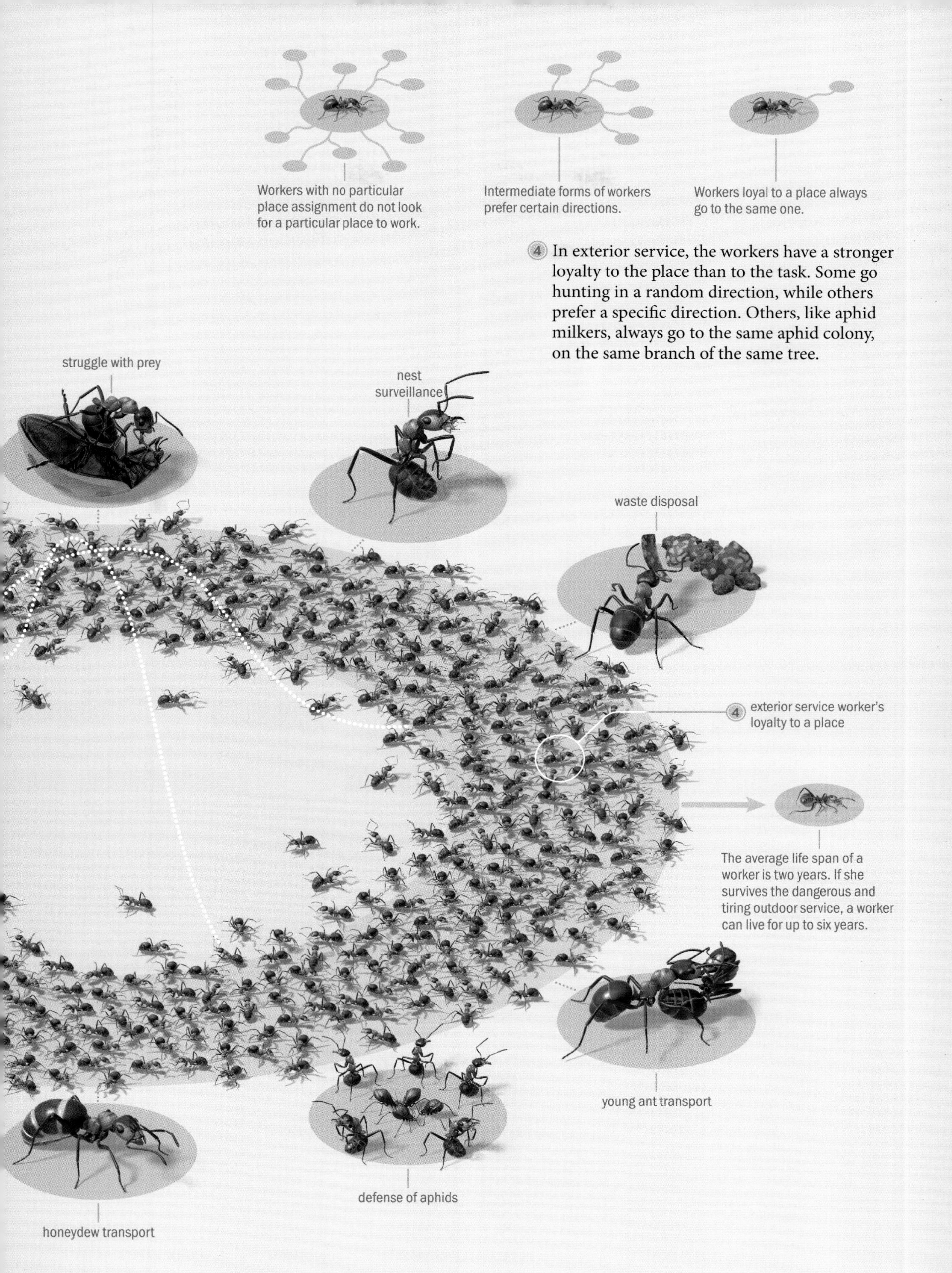

4 In exterior service, the workers have a stronger loyalty to the place than to the task. Some go hunting in a random direction, while others prefer a specific direction. Others, like aphid milkers, always go to the same aphid colony, on the same branch of the same tree.

8.1 The attack

An aggressor has opened the anthill dome. Suddenly, daylight floods the darkened rooms inside the nest. This unexpected disturbance causes the colony to get excited.

The workers run in all directions to draw their congeners' attention to the threat. Whenever they encounter another worker, they bump or pat it with their antennae.

The general excitement is amplified by the combination of formic acid that the ants project around them and the secretions of their Dufour's glands. This mixture's odor acts as an alarm signal and attracts part of the colony to the place of its diffusion.

Ants react differently to the disturbance and to the alarm signal. The young workers, whose poison gland is still inactive, behave defensively and take refuge in the nest's dark depths.
Older workers, whose poison gland is active and full, become aggressive. Ready to fight, they come out of the anthill to locate the cause of the disturbance.

Wild boars are ransacking the anthill. In search of food, they scavenge the ground and the dome for ant and beetle larvae.
The boars unearth a beetle larva cocoon. The beetle lays its eggs on the domes of anthills. When they hatch, the "white grubs" burrow into the dome and make a cocoon of saliva, plant material, and soil. They feed on plant matter they find in the dome. The large grubs are very nutritious and are prized by wild boars and other forest animals.

Several nurseries full of larvae are ripped open by the attackers. The workers flee, carrying the larvae to safety.

8.2 Defense of the nest

A young wild boar has found a white grub and is devouring it. Meanwhile, the workers collectively attack the boars and try to push them back.

Some of the workers throw themselves into body-to-body combat. Climbing on the boars in search of vulnerable areas devoid of fur, they bite the skin and project formic acid into the wounds.

Another group of workers attacks from a distance. Assuming a threatening posture, they project formic acid on the boars from several centimeters away. The strong odor of formic acid draws other workers into the struggle.

Formic acid has a pungent odor
and irritates the eyes and the nasal
and oral mucous membranes. The
effect is multiplied by the large
numbers of ants.

Although the ants are not
able to seriously harm the
boars, the effect is unpleasant
enough to drive them away.
While some of the ants are
fighting, the others begin to repair
the nest and collect the materials
scattered by the attackers.

8.3 Exploration of the surroundings

The wild boar attack caused great damage to the anthill. The ants must now restore it or find another location.

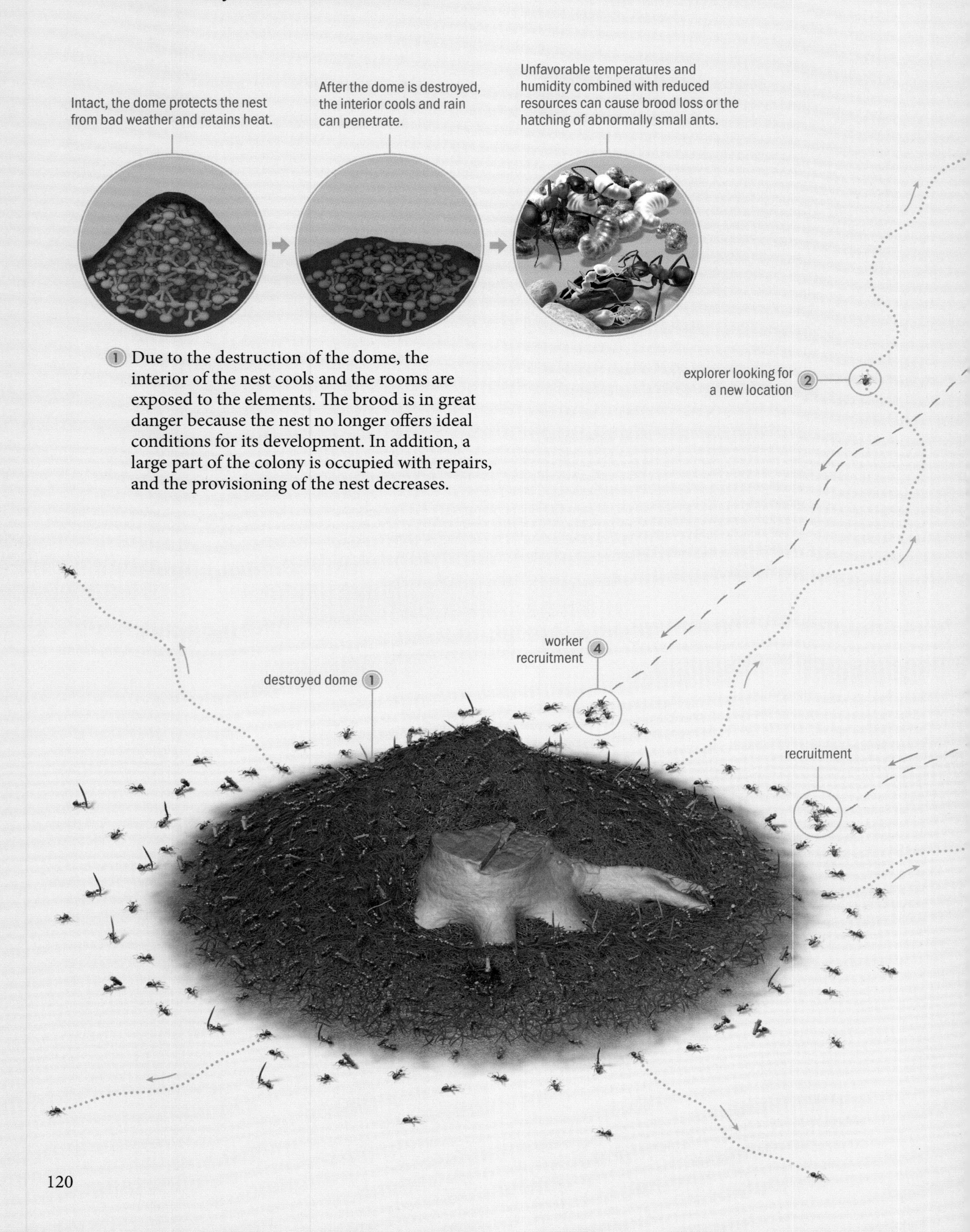

1 Due to the destruction of the dome, the interior of the nest cools and the rooms are exposed to the elements. The brood is in great danger because the nest no longer offers ideal conditions for its development. In addition, a large part of the colony is occupied with repairs, and the provisioning of the nest decreases.

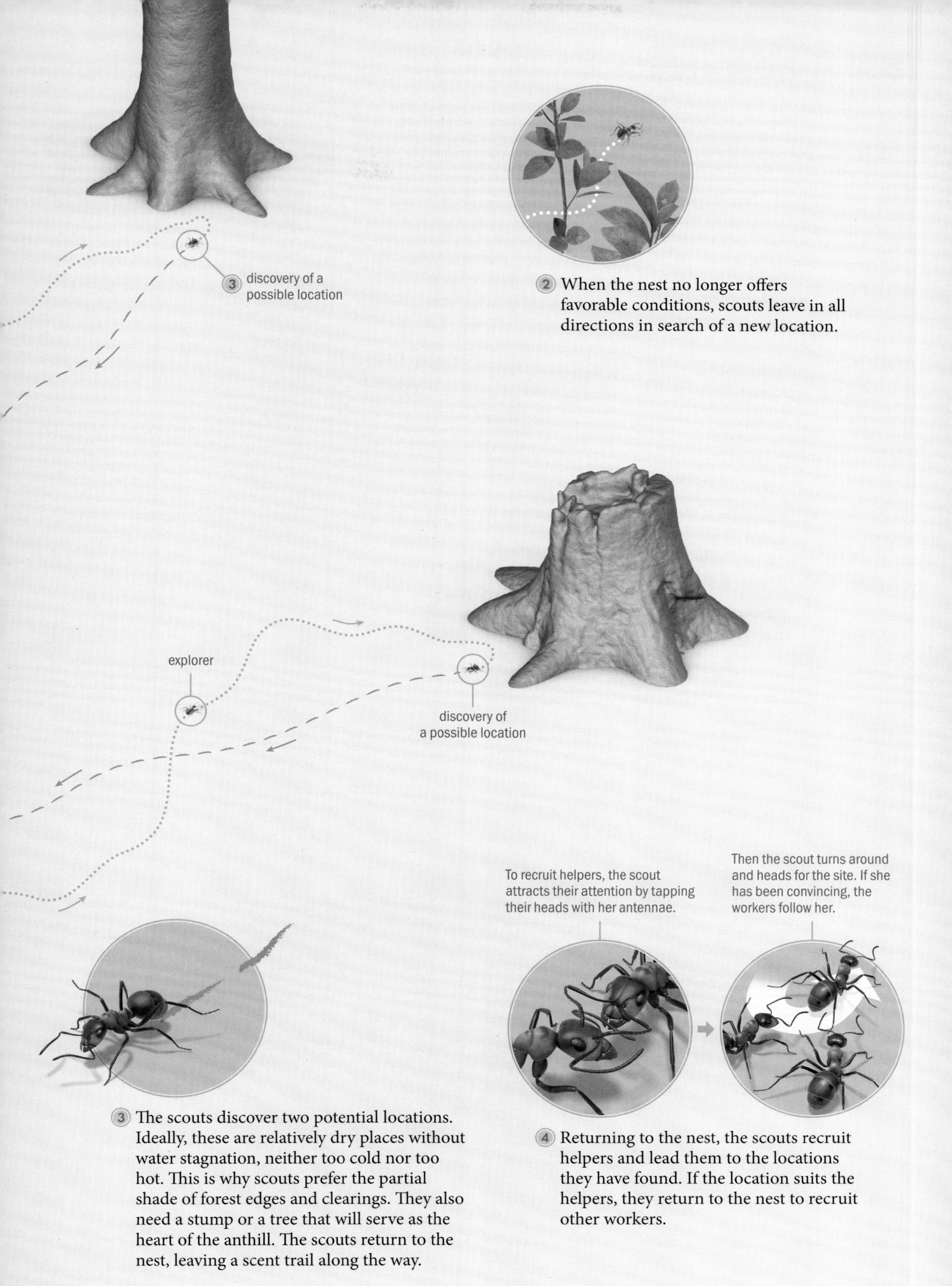

2 When the nest no longer offers favorable conditions, scouts leave in all directions in search of a new location.

3 The scouts discover two potential locations. Ideally, these are relatively dry places without water stagnation, neither too cold nor too hot. This is why scouts prefer the partial shade of forest edges and clearings. They also need a stump or a tree that will serve as the heart of the anthill. The scouts return to the nest, leaving a scent trail along the way.

4 Returning to the nest, the scouts recruit helpers and lead them to the locations they have found. If the location suits the helpers, they return to the nest to recruit other workers.

8.4 A new nest

The workers prepare for the construction of nests in the new locations. A partial move of the ant colony begins at both sites.

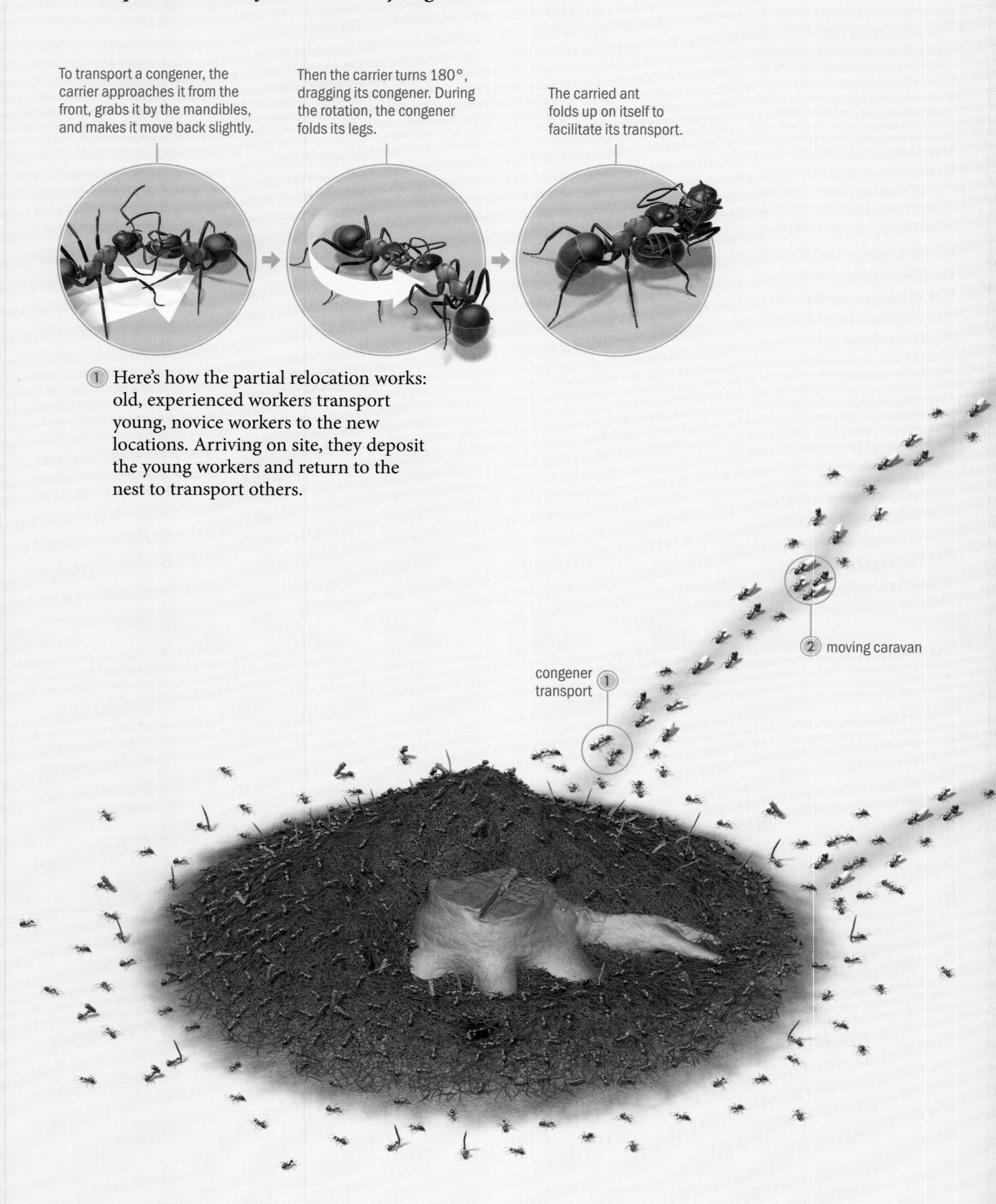

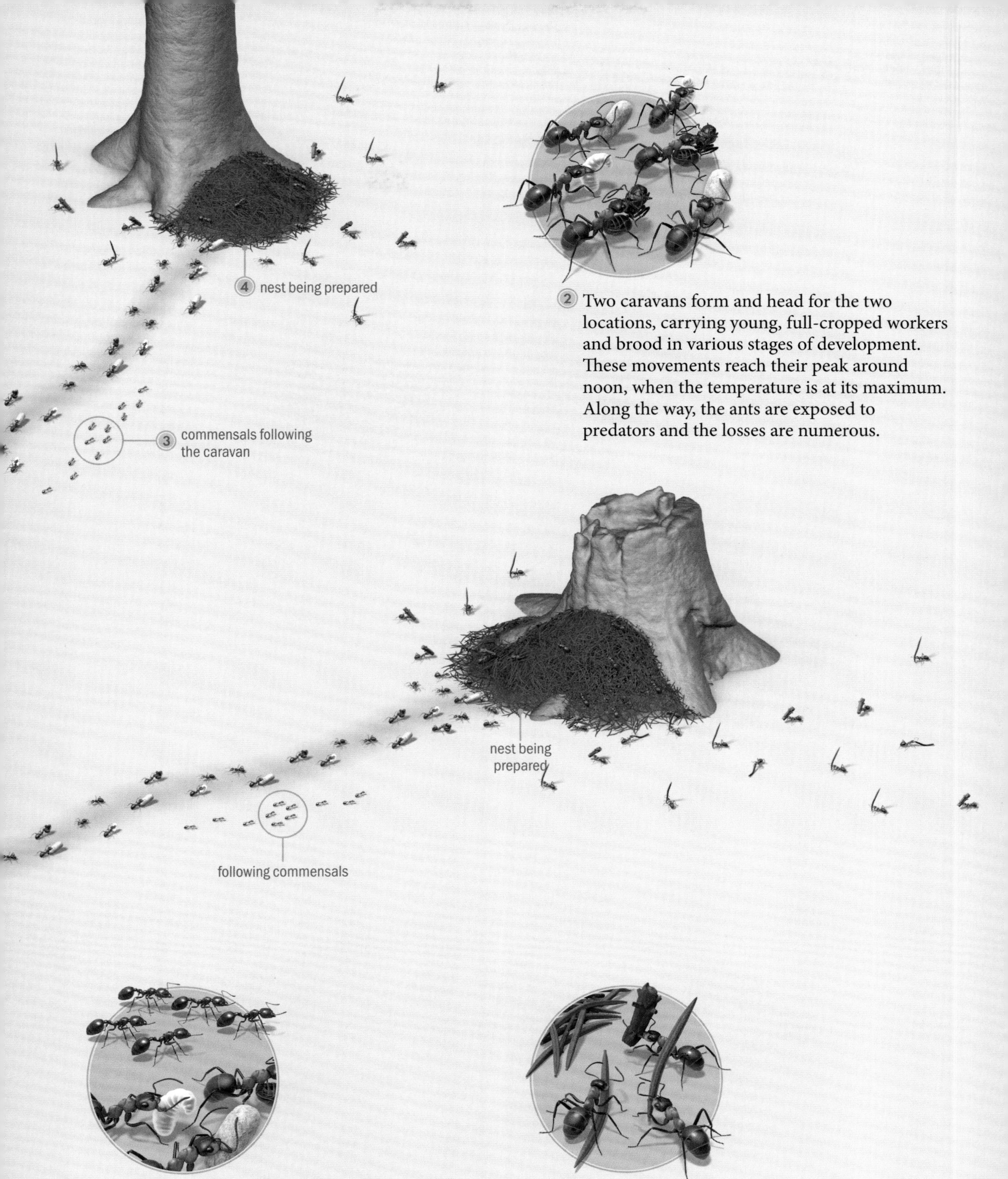

② Two caravans form and head for the two locations, carrying young, full-cropped workers and brood in various stages of development. These movements reach their peak around noon, when the temperature is at its maximum. Along the way, the ants are exposed to predators and the losses are numerous.

③ Some commensals of red wood ants, such as the host ant *Formicoxenus nitudulus*, are able to perceive ant scent trails. They follow its caravans to the new nests.

④ The number of workers increases in both places. They prepare for the construction of new nests, digging underground galleries and bringing materials for the dome.

8.5 Colony relocation

A red wood ant colony has only one queen. She must therefore choose between the two locations. Once the decision is made, the construction of the nest begins in the chosen site.

① The population does not grow at the same rate in the two locations: one of them is more popular and attracts more workers. As soon as the population at that location exceeds a certain threshold, the queen is transported there.

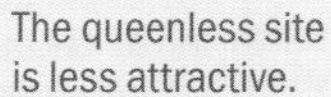

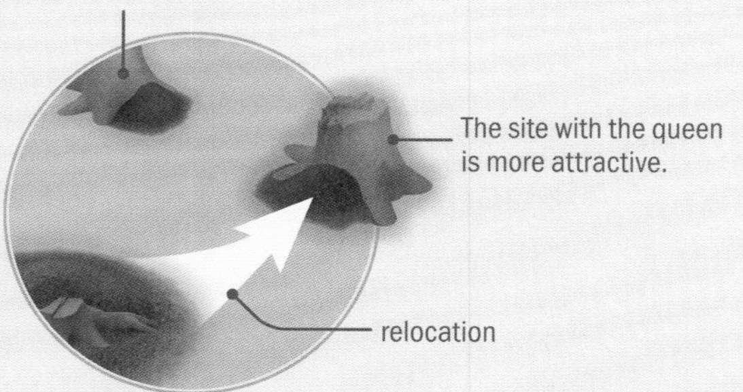

② The queen's presence and scent increase the site's attractiveness. The relocation switches to this site, and the entire colony goes there in a long procession.

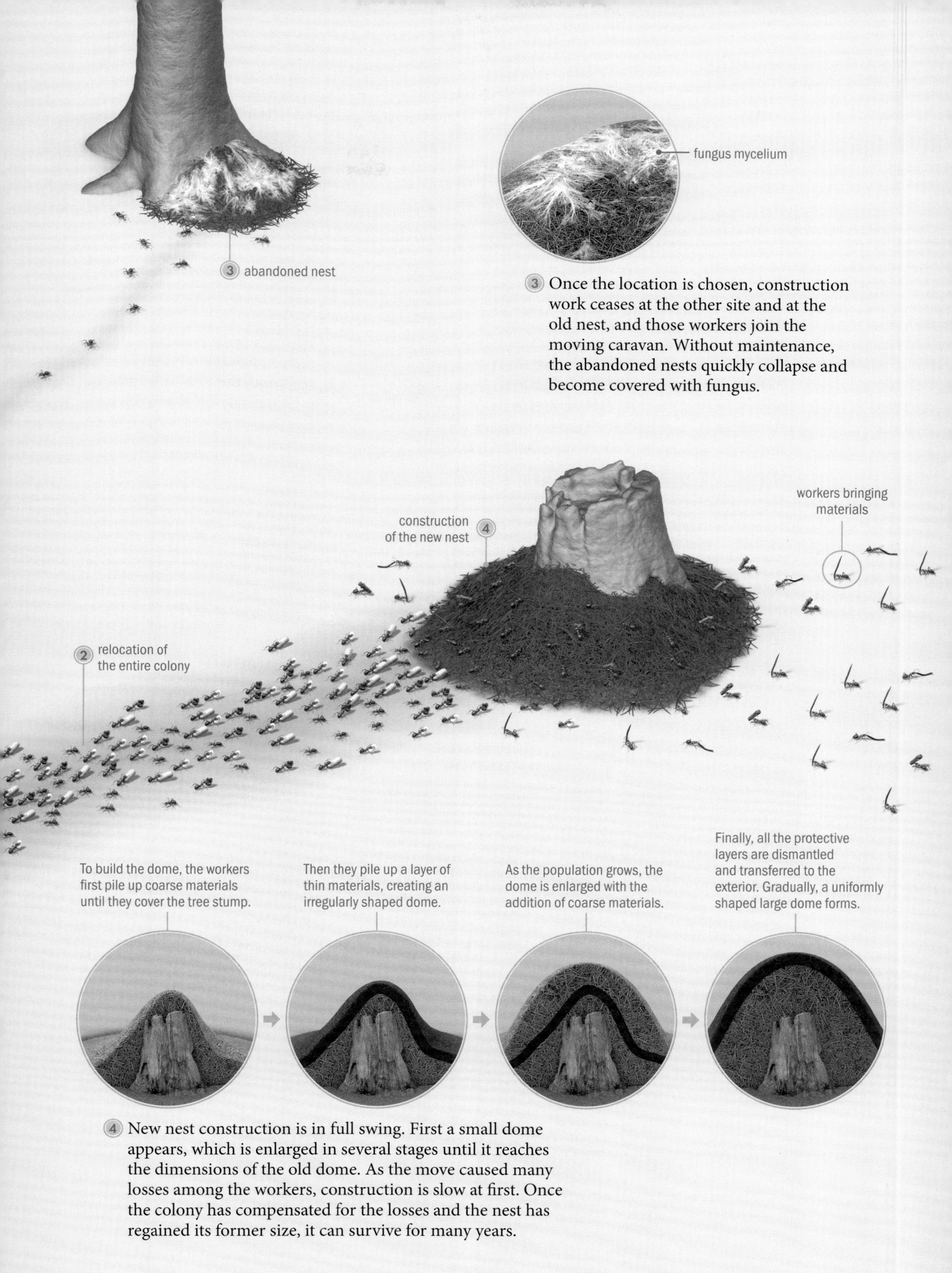

3 Once the location is chosen, construction work ceases at the other site and at the old nest, and those workers join the moving caravan. Without maintenance, the abandoned nests quickly collapse and become covered with fungus.

4 New nest construction is in full swing. First a small dome appears, which is enlarged in several stages until it reaches the dimensions of the old dome. As the move caused many losses among the workers, construction is slow at first. Once the colony has compensated for the losses and the nest has regained its former size, it can survive for many years.

The author

Armin Schieb works as a freelance illustrator in Hamburg, Germany, and has a special interest in scientific, technical, and fantastical subjects. He studied illustration at the Hamburg University of Applied Sciences, where the idea for this book arose as part of his master's project.

Initially, his project was focused on the subject of collective intelligence. During his research, he discovered in an issue of *National Geographic* a photographic report on the army ants of South America. The ants were shown up close, busy with their social activities. He was so intrigued by their diverse social life, their cooperation in solving complex tasks, and their interaction with the environment that he decided to illustrate collective intelligence through ants.

He chose the red wood ant because he could observe it directly in nature and gather material for his work. Using his sketches, photos he took and collected himself, and scientific drawings, Schieb created 3-D digital models of ants, their nests, and other images, which he combined into an ambient setting with realistic lighting.

One of his concerns during the conception of the book was to represent the ants and their environment in a realistic way, in order to familiarize the reader with these discreet and little known animals. For him, it was important that ants be perceived as appealing creatures, and he therefore avoided depicting frightening scenes. His wish: that the success of the book contribute to awakening the reader's interest and respect for ants and that the reader consider them worthy of protection.

Original title: *Das Ameisenkollektiv*.

Published by Princeton University Press
41 William Street, Princeton, New Jersey 08540
99 Banbury Road, Oxford OX2 6JX

press.princeton.edu

ISBN 9780691255927
ISBN (e-book) 9780691255934

British Library Cataloging-in-Publication Data is available

Editorial: Robert Kirk and Megan Mendonça
Production Editorial: Natalie Baan
Text Design for English Edition: D & N Publishing, Wiltshire, UK
Jacket Design: Wanda España
Production: Ruthie Rosenstock
Publicity: Caitlyn Robson and Matthew Taylor
Copyeditor: Jennifer McClain

Jacket image: Armin Schieb

This book has been composed in Franklin Gothic URW and Minion

Printed on acid-free paper. ∞

Printed in Slovakia

10 9 8 7 6 5 4 3 2 1